Westover

Man, Base and Mission

Frank Faulkner

Hungry Hill Press
Springfield, Massachusetts

"Westover always left more positive thoughts in my memory than negative thoughts," said Carol Bernier-Gawle, who was a child when paper "bombs" struck her Chicopee neighborhood, and later witnessed 50 years of change at Westover. "In spite of everything, I think life would not have been the same -- my life would not have been the same -- without Westover having been there. I think it gave me a broader outlook. Westover made me more aware, both as a child and as an adult, that the world was a whole lot bigger than just the city of Chicopee or the town of Easthampton....I'd have to say it made me aware that what we did here affected a lot more people than just the people living here. It affected people in a lot of places that I'd never seen -- places I would never see, or never know about -- but I think that awareness was created in me just by the fact that Westover was there."

First Printing

International Standard Book Number:
0-9616486-1-9

Library of Congress Catalog Card Number

Printed in the United States of America

Dedicated to the many hundreds of young airmen for whom Westover Field, and later Westover AFB, was the last American soil upon which they ever walked.

Preface

Researching the fascinating history of Westover Field brought back memories of growing up in Springfield and watching the television roll when a B-52 roared over, or stopping dinner conversation when a water-injected KC-135 dominated our home with a sound like yards of canvas being torn only an inch from our ears.

From our back porch in the Six Corners neighborhood, I began watching C-54s and later B-36s coming off Runway 23. About the time the B-52s arrived, we had moved to the Pine Point section and my view of KC-97s, B-47s, KC-135s and B-52s improved. I was 13 and sleeping in Dick Mayer's backyard in Sixteen Acres when a KC-135 plowed across the Mass Turnpike and exploded in the Machowskis' backyard. We heard the rolling explosion six miles away, and for nearly five minutes the gigantic fireball turned the night into a bright red sunset. We assumed an atomic bomb had hit Westover and life in Springfield would end within minutes. I don't remember being frightened. We were resigned to our fate, because we had been raised knowing Westover AFB was one of the most important bases in the world.

My first visit to Westover AFB came as a Civil Air Patrol cadet directing traffic on Armed Forces Day 1959. Vin Blanchard's father photographed us in front of an F-86. Through CAP, I learned to fly an L-4 Grasshopper before I knew how to drive a car. In May 1961, I crash landed at Westover with Helmuth Lehburgher, and SAC security police with carbines and BARs surrounded our L-16.

The Vietnam War and work overseas separated me from the area until I moved to Granby in the early 1970s. After SAC left, deer living near the nuclear storage facility at Stony Brook began peering in windows along New Ludlow Road. Dangerously tame, the deer had been fed by GIs, and lights meant chow time. Westover, like the deer, craved assistance.

Fields in Granby provided the setting for cheap dates as people parked in the late afternoons and evenings to watch the Air Force Reserve C-123s and C-130s make CDS and troop drops on Westover's DZ.

The coming of the C-5As brought protests from a few families in the town, but Greater Springfield residents overwhelmingly supported the international mission of the huge and loud Galaxies. The renewed interest in Westover was demonstrated clearly in June 1987 when more than 350,000 people "voted with their feet" and came to the base for an airshow and outpouring of support for the Westover mission.

The massive public interest sparked my interest in researching the history of Westover and the pilot for whom the base was named. But no history of an installation is accurate without the sociopolitical context and even the psychohistorical prompting that developed the various missions of the base.

I take full blame for errors in judgment in assessing the mission development and the accuracy of newspaper clippings, official reports and the memories of others. There were many conflicting dates, numbers and names. I did my best to judge what should be included on these pages.

I am indebted to Rick Dyer and Gordon Newell of the 439th MAW public affairs staff for permission to publish official USAF and USAAF photographs. All photos are Air Force unless otherwise noted. Wayne Phaneuf, managing editor of the Springfield *Union-News*, aided this project with many file photos. Chief Robert Adams provided valuable assistance with historical photographs and his wealth of knowledge about the history of Westover AFB. Filmmaker John D. Gordon provided transcripts from the outtakes of his excellent video documentary *Fifty Years at Westover*. John J. Keough and Art Monahan of the Galaxy Community Council have given tremendous support to this project, and a substantial portion of the cover price has been donated to the GCC for the men and women of Westover.

Special thanks is given to Jeff Spear, Bob Catalanotti and John Tabb of John C. Otto Company Inc., commercial printers. Sincere appreciation is extended to Christine Mora for the many hours spent on the design of this book, and to Dr. Jim O'Connell of Hungry Hill Press for production of this history.

And I am deeply grateful to my wife, Melinda Phelps, and daughters, Maura and Kathleen, for tolerating the evenings and weekends devoted to this lengthly project.

FRANK FAULKNER
Dingle, County Kerry
June 6, 1990

Anthony J. Stonina
Mayor of Chicopee, Massachusetts

THE NORTHEAST AIR BASE

An international dream comes to Western Massachusetts

The dream of having an international airfield in Chicopee to serve Western Massachusetts emerged in 1926, during the "roaring Twenties" and only 23 years after the Wright Brothers' first flight.

That first flight in 1903, incidently, was a foot shorter than the cargo hold in the C-5A Galaxies assigned today to Westover Air Force Base. The aviation dream of Western Massachusetts took many turns before the last decade of the century when the 439th Military Airlift Wing was operating 16 huge C-5As from this important airfield.

Westover AFB is the nation's largest Air Force Reserve base and more than 2,400 men and women from the cities and hamlets of New England are members of the many units on this 2,300-acre base.

Westover has always been important. For a half century, this base has served the needs of America. The wide and long runways are among the best on the East Coast.

The concept of turning Chicopee's tobacco fields into an airfield came during a golden post-war era when the nation was prospering and the future of Western Massachusetts was very bright.

The vision of a major Greater Springfield airport here was promoted by Anthony J. Stonina, a young Polish immigrant and first-term Chicopee alderman.

MAYOR ANTHONY J. STONINA

The airfield idea became planted in Stonina's head during 1926. "A retired Army Air Corps instructor named Will Sergeant had been taking neighbors up for rides in an old war surplus plane he had," Stonina said. "One day he came to me and told me Chicopee had the most ideal spot in the East for a municipal airport. Not only was it physically ideal, but with the population and power supplies so near, it should be looked into. So I took a ride up over the valley with him and sure enough, he was right. But when I suggested the idea to the mayor at our next meeting, he said it would be too expensive and extravagant at the time. Nevertheless, the idea stayed with me."

Faded newspaper accounts from the Roaring '20s indicate Stonina could not convince local officials of the soundness of his idea to turn the flat tobacco plains into an airport. Local politicians continued to say the concept of air travel was too extravagant for a small industrial city.

When the stock market crashed in the autumn of 1929, Stonina was three years into his political career and a mayoral candidate. He won the election and tried to save Chicopee from slipping into bankruptcy.

Stonina's leadership was appreciated. He even managed to coax city workers during one period to labor without pay while he sought funding for Chicopee.

Ben Mathis, a former B-24 pilot at Westover Field and a longtime friend of Stonina, remembers tales of how the mayor worked like "the dogged salesman he was" to convince Army Air Corps officials to choose the tobacco fields for an air base.

Mathis had dated the mayor's daughter while he was a young Air Corps captain at Mitchell Field in New York. He was later transferred to Westover and married her. He

described Stonina as "probably the most patriotic individual I've ever known."

Stonina had come to America due to misfortune. His father had been working in Northampton during 1908 to earn cash to buy the land his wife and six young children were tending in Austria-Hungary. One afternoon young Anthony Stonina and his mother came out of the farm fields and saw their thatched roof cottage burning. Soon, their crops were afire.

Mrs. Stonina sent her children off to America.

Calvin Coolidge was a mayoral candidate in Northampton when Anthony Stonina arrived in 1909. The young Polish-speaking immigrant remembered being fascinated by the posters, parades and political rallies. The impression remained with Stonina. So did his Republican leanings and loyalty to Mayor, then Governor, and finally President Coolidge. When Calvin Coolidge died in 1933, Mayor Stonina went to Washington as an honorary pallbearer.

His father moved the family to Chicopee in 1910 and opened the city's first gasoline station. Nine years later, Stonina and his brothers began selling automobiles at the Chicopee Tire and Accessory Store.

Stonina needed the store after he had become mayor and was forced in 1933 to stop paying city employees. He had

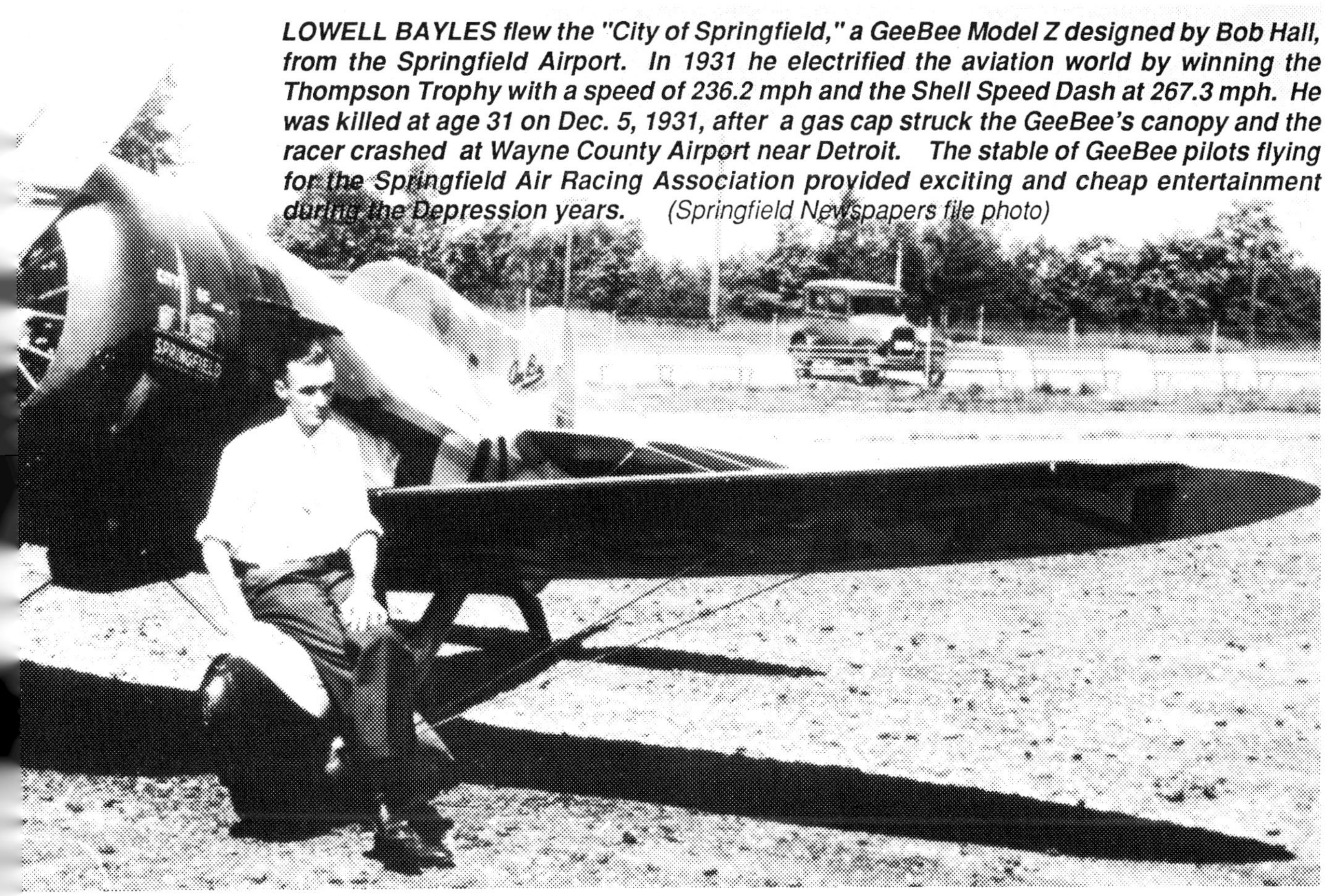

LOWELL BAYLES flew the "City of Springfield," a GeeBee Model Z designed by Bob Hall, from the Springfield Airport. In 1931 he electrified the aviation world by winning the Thompson Trophy with a speed of 236.2 mph and the Shell Speed Dash at 267.3 mph. He was killed at age 31 on Dec. 5, 1931, after a gas cap struck the GeeBee's canopy and the racer crashed at Wayne County Airport near Detroit. The stable of GeeBee pilots flying for the Springfield Air Racing Association provided exciting and cheap entertainment during the Depression years. *(Springfield Newspapers file photo)*

hoped to establish credit during the Depression and get a loan for the city. After two months without success, he slashed the Chicopee budget 15 percent, and dropped his own salary to $2,500 a year. Two months later, two banks loaned Chicopee $1.4 million and the city was back on its feet. But the loan came at the end of Stonina's first term, and he was trounced at the polls during the November election.

As the Depression was abating in 1935, Stonina regrouped his political and financial forces, and became the first man to win back the mayor's job in Chicopee.

And he kept refining his idea of an international airfield in Chicopee.

That same year, Congressional leaders spied the gathering war clouds in Europe and passed the Army Air Base Bill, better known as the "Wilcox Act," calling for construction of several airfields deemed vital to the security of the nation. One major military airfield was to be in the Northeast.

In late 1938, General Henry "Hap" Arnold became Air Corps chief of staff after his friend, Major General Oscar Westover, died in a plane crash. As one of his first acts, Arnold recommended the War Department appoint a board of officers to select sites for the modern airfields authorized years before by the Wilcox Act.

By then, Mayor Stonina was in his fourth term in Chicopee's city hall.

When he learned of Arnold's recommendation, Stonina fired off a telegram on Nov. 22, 1938, to the Secretary of War suggesting Chicopee as a suitable site for the Northeast Air

MAUDE TAIT MORIARTY was one of the many local pilots who captured the imagination of local aviation enthusiasts. Her father owned a dairy farm where the Springfield Shopping Plaza is now located. James Tait developed the Springfield Airport and invited the famous Granville brothers to set up an aircraft manufacturing operation in Springfield. Maude Tait was the first licensed female pilot in Massachusetts. She set a world speed record in a GeeBee Model Y Sportster during the 1931 air races in Cleveland. After she married U.S. Commissioner James Moriarty, she became active in promoting aviation in the Pioneer Valley and establishing the Civil Air Patrol. *(Photo courtesy of Melinda Phelps)*

Base. This effort was followed by a substantial written report from the city's 10-man aviation committee on soil data, land prices, proximity to Greater Springfield industrial centers, weather information and diagrams of the strategic location. The committee collected photographs of the area and sent the entire package to the War Department. The city's effort was aided by U.S. Rep. Charles R. Clason of Springfield, who urged Army generals to consider the tobacco plains because the variation in elevation at the proposed airfield site was only one foot.

In July 1939, a contingent of five Army officers visited the Pioneer Valley to tour the Springfield Airport, where the GeeBee air racers were born, to examine Bowles Airport in Agawam and the airport in Westfield.

Stonina learned about the Army contingent during a telephone call from *Springfield Union* reporter Dick Bailey. Stonina rounded up automobiles from his dealership and headed to Springfield. He met with the officers and took them on a guided tour to survey the flat Chicopee tobacco plains which extended into Ludlow. He sold them on the tobacco flats.

A truck convoy of Army engineers arrived from Alabama on Aug. 8, 1939, to begin survey work on the flat area of Chicopee and Ludlow. The engineers were billeted at a former Civilian Conservation Corps camp in Westfield.

In the meantime, Stonina petitioned the state Legislature through Senator Chester Skibinski of Chicopee to clear obstacles for the federal government taking the land.

But the whole effort changed in September.

"All the News That's Fit to Print."
The New York Times.
EXTRA
NEW YORK, FRIDAY, SEPTEMBER 1, 1939.
THREE CENTS
GERMAN ARMY ATTACKS POLAND;
CITIES BOMBED, PORT BLOCKADED;
DANZIG IS ACCEPTED INTO REICH
Bulletins on Europe's Conflict
HOSTILITIES BEGUN
Warsaw Reports German Offensive Moving on Three Objectives
FREE CITY IS SEIZED
Forster Notifies Hitler of Order Putting Danzig Into the Reich
Hitler Acts Against Poland
HITLER GIVES WORD
In a Proclamation He Accuses Warsaw of Appeal to Arms
FOREIGNERS ARE WARNED
They Remain in Poland at Own Risk—Nazis to Shoot at Any Planes Flying Over Reich
By OTTO D. TOLISCHUS

CHAMBERLAIN ANNOUNCES BRITAIN IS AT WAR WITH GERMANY
"All the News That's Fit to Print."
The New York Times.
EXTRA
NEW YORK, SUNDAY, SEPTEMBER 3, 1939.
TEN CENTS
Section 1
BRITAIN AND FRANCE IN WAR AT 6 A. M.;
HITLER WON'T HALT ATTACK ON POLES;
CHAMBERLAIN CALLS EMPIRE TO ...
SOVIET IN WARNING

On Sept. 1, 1939, Nazi forces smashed into Poland and the European war was ignited. Stonina was angered by the Nazi invasion of his native land and rushed to Washington, D.C., to meet with Arnold and to push for construction of the Northeast Air Base.

Within only two weeks -- on Sept. 15, 1939 -- Secretary of War Harry H. Woodring announced Chicopee would "become an important link in the chain of East Coast defense" because the tobacco plains had been selected as the site for the Northeast Air Base.

A short while later, $2,070,918 was allocated by the War Department to purchase a 7.5 square-mile plot for the Northeast Air Base.

On Sept. 21, the War Department took official steps to acquire title to the land. The first construction equipment arrived on Oct. 2 and on Oct. 18 the federal government acquired the land as U.S. Attorney Edmund J. Brandon filed an order of seizure by eminent domain and deposited $207,091.80 with the federal court in Boston.

[Chap. 463]

An Act granting the consent of the commonwealth to the acquisition by the United States of America of certain lands in the city of Chicopee and the town of Ludlow for the purposes of an army air-base.

Be it enacted, etc., as follows:

Section 1. The consent of the commonwealth is hereby granted to the acquisition, on or before December thirty-first, nineteen hundred and forty, by the United States of America, by purchase or condemnation, for the purposes of a United States army air-base, of approximately eight square miles of land located in the northeast corner of the city of Chicopee and the northwest corner of the town of Ludlow; provided, that a suitable plan of the land so acquired has been or shall be filed in the office of the state secretary within one year after the acquisition thereof.

Section 2. Jurisdiction over said land is hereby granted and ceded to the United States of America, but upon the express condition that the commonwealth shall retain concurrent jurisdiction with the United States of America in and over the land so acquired, in so far that all civil processes, and such criminal processes as may issue under the authority of the commonwealth against any person or persons charged with crimes committed without said land and all processes for the collection of taxes levied under authority of the laws of the commonwealth, including the service of warrants, may be executed thereon in the same manner as though this consent and cession had not been granted; and exclusive jurisdiction over any such land or any part thereof shall revest in the commonwealth whenever it shall cease to be used by the United States for the purposes specified in section one.

Approved August 12, 1939.

Stonina convinced the Boston & Maine Railroad to build an extension to the base.

In early November more than 200 homes and land owners in the condemned base area were informed they would have to vacate their property within 30 days.

Mayor Stonina had a tough time convincing many of Chicopee's Polish farmers that the taking of their fertile land was in the best interests of the nation, their community and their own future. He was voted out of office in November after four terms as Chicopee mayor. He became a car salesman once again.

Stonina later recounted the end of his political career. "City fathers at the time were afraid 10,000 new men in a community would corrupt the city's morals and we'd have to have a curfew," Stonina said. "Then they worried about bombers crashing into their homes at night and the cost of educating all the personnel's children in the public schools. Finally, when someone got up on a soapbox and told them that the government's $2 million price tag on the area was too low, I felt the bottom drop out." Stonina began selling cars as "A.J. Stonina Inc." of Chicopee.

Later that month, President Franklin D. Roosevelt signed a $750,000 WPA project bill for the airbase and construction bids were solicited in January 1940.

General Arnold announced on Dec. 1, 1939, that the air field would be designated Westover Field to honor the memory of his pal and long-time boss, Maj. Gen. Oscar Westover, the former Army private who had learned to fly.

MAJOR GENERAL OSCAR WESTOVER

Visionary pioneer of the Army Air Corps

Maj. Gen. Oscar Westover never saw the "Northeast Air Base" which was carved out of lush Western Massachusetts tobacco fields and named in his memory on April 6, 1940.

But Westover -- the Army Air Corps Chief who was nicknamed "Tubby" and regarded as America's "Number One Flier" -- undoubtedly would have liked the Chicopee airfield.

"He loved to fly," said his daughter, Patricia Freudenthal, who resides today in Elkwood, Va. "He just felt like he was soaring with God."

The two-star general, who died Sept. 21, 1938, at age 55, in a fiery airplane crash near Burbank, Calif., was a pioneer of modern military aviation.

An Army private who won an appointment to West Point, Westover was regarded as a pilot's pilot. He was also a visionary who early on realized the impact which the airplane would have on warfare in the 20th Century.

Although he served as its chief for just 636 days, Westover helped found the fledgling Army Air Corps. He also helped to prepare America for the world war which he never lived to see.

In an editorial tribute published shortly after his death, a Washington, D.C., newspaper wrote that General Westover "devoted his life to the work of building up a powerful aerial defense for his country."

President Roosevelt lauded him as a "gallant soldier" and "a true friend."

The legendary General H.H. "Hap" Arnold, Westover's successor as head of the Air Corps, was the one who suggested the massive military air base then under construction in Massachusetts between Ludlow and Chicopee be named in honor of the fallen aviator.

Thousands of GIs would come to know Westover -- the air base -- during the next half century.

But few of them knew much about Westover, the man.

Oscar Westover was a Michigan native, born in Bay City on July 23, 1883. He enlisted in the Army in 1901 and entered the United States Military Academy the following year.

Fellow plebes at West Point dubbed the five-foot five-inch future general "Legs," "Shorty," "Eastunder," and "Tubby." But they also held him in high esteem.

While at West Point, Westover was a cadet officer, football quarterback, standout gymnast, and YMCA president in the class of 1906.

CADET OSCAR WESTOVER, 1906

A classmate, Forrest E. Williford, later wrote that his friend Oscar was a "happy blend of athlete, soldier, Christian and gentleman."

Following his graduation and commissioning in 1906, Westover was trained at the Presidio of Monterey and served as an infantry officer at Vancouver Barracks in Washington.

While assigned in Washington, he married Adelaide Regina Bainbridge on Dec. 31, 1907. Shortly thereafter, the young lieutenant and his bride were transferred to the Philippines where he served with the 14th Infantry Regiment.

Adelaide Regina Bainbridge Westover was a direct descendant of Commodore William Bainbridge, a naval hero who commanded the *U.S.S. Constitution* during the War of 1812.

In 1911, Lieutenant Westover returned to West Point as an instructor in the academy's Department of Drawing.

Following an assignment in Alaska in 1914, he returned again to the Military Academy as a captain and assistant professor of drawing.

As the winds of war blew through Europe, Westover was assigned to the Army's Signal Corps and given his first real introduction to newfangled invention called the airplane.

His assignment during World War I was with the Signal Corps bureau of aircraft production. He quickly became a temporary lieutenant colonel and -- because early airframes were built of wood -- "an expert in the handling of spruce forest and mills in the Pacific Northwest."

For his work during the war, overseeing construction of the British-designed but American-built de Havilland 4 aircraft, Westover was awarded the Distinguished Service Medal.

When the Armistice was signed, Westover was a temporary colonel and executive officer with the Signal Corps' air service.

The naval connection

When Lieutenant Oscar Westover married Adelaide Bainbridge on Dec. 31, 1907, he linked his family to a famous naval tradition.

She was a direct descendant of Commodore William Bainbridge, a U.S. Navy officer who captained the *USS Constitution.*

Commodore Bainbridge was at the helm of "Old Ironsides" when the famous warship captured the British frigate "Java" off the coast of Brazil during the War of 1812.

Commodore Bainbridge had also fought against the Barbary pirates and was in command of the frigate "Philadelphia" when she was captured by the Tripolitans in 1803. He was imprisoned for a time in what is today Libya.

The Westovers' son, Charles Bainbridge Westover, was nicknamed "Tubby Two" and became a three-star Air Force general.

ADELAIDE BAINBRIDGE WESTOVER

COLONEL WESTOVER and an unidentifed Army Air Corps captain about 1930. Westover later became a trustee of the National Geographic Society. (Freudenthal Collection)

Intrigued with the boundless potential of aviation and thoroughly convinced of its military applications, "Tubby" Westover had a new goal -- he wanted to fly.

At age 38, when military colleagues just slightly older were eying retirement, Westover eyed the skies.

In 1921, after reverting to his permanent grade of major, he enrolled in the Army Balloon School at Omaha, Neb. The following year he learned to pilot cigar-shaped dirigibles at the Army Airship School at Langley, Va.

By 1923, the 40-year-old officer had earned his silver pilot's wings from the Air Corp's Primary Flying School at Brooks Field and his advanced rating from the Advanced Flying School at Langley.

At the time, he was one of only three or four Air Corps officers holding every Army pilot rating -- balloon observer, dirigible pilot, aerial observer and airplane pilot.

As a balloonist, Westover first achieved international fame.

On June 3, 1922, he won a national championship and shattered a world speed record in a balloon race from Milwaukee, Wisc., to the St. John's River in Quebec, Canada.

Westover's Army balloon traveled the 911-mile distance in less than 18 hours. His record-setting pace was aided greatly by some abnormal gale force winds exceeding 80 miles-per-hour.

The *Washington Post* reported that during the race, "the atmosphere was so heavy that (Westover) and his assistants could not see the landscape below."

"In our attempt to make a record flight, we threw overboard our camping outfit, our radio set and all our emergency paraphernalia," the victorious Westover told reporters.

According to news accounts, his daughter Patricia -- then four years old -- and her seven-year-old brother Charles were at home in Washington, D.C., "eagerly waiting for the news" of their father's triumph.

Patricia Westover later married an Army pilot, Col. Wilhelm Freudenthal. Her brother, Charles, graduated from West Point in 1937 and later became an Air Corps pilot and three-star Air Force general who served as vice-commander of the Tactical Air Command.

Yellowing newspaper clippings aren't Mrs. Freudenthal's only keepsake from her father's balloon adventures. She also has a momento from a different -- and less successful -- race which Oscar Westover entered.

In August, 1922, just two months after the record-setting trek to Canada, Westover and co-pilot Capt. Carleton Bond entered the international Gordon Bennett Cup competition.

The race started in Geneva, Switzerland, and the two Army officers were in the lead as they floated over the Alps.

"I wish you could have had the thrill of flying above the Alps on the night we made that flight," Westover told a National Geographic Society audience.

"Those of us who were ahead of the storm had broken clouds ahead, moonlight, starlight and snowcapped mountains," he said. "Down beneath us were the shimmering valleys, the lights, the streams and also the tinsel lights of the villages."

The race, however, ended abruptly for Westover and Bond. As their lighter-than-aircraft floated low over a Hungarian village, excited townspeople gathered below. Fearing the balloon was in trouble, the villagers grabbed the dangling tether rope and pulled the men to earth. The unlucky leaders were out of the race.

Mrs. Freudenthal still has a copy of a letter written by three Hungarian officers describing her father's plight. They wrote how "people rushed on and held seize of the rope, not understanding the shouts of Major Westover to let it go." The letter concluded: "Well understanding that they will have to explain the reasons of their landing, we give this statement of our own accord and are sorry for the Stars and Stripes."

(Freudenthal Collection)

Perhaps the funniest anecdote about General Westover concerned an incident which happened even before he won his Army pilot's wings.

During World War I, Westover had just returned to his duty station in Washington, D.C., following an inspection trip to the West Coast.

When an assistant arranged for him to take an orientation flight in a military bi-winged plane at nearby Langley Field in Virginia, Westover agreed eagerly.

Prior to takeoff, Westover mentioned to his pilot how he "had a fine flight out West."

After the two men were airborne, the pilot yelled for Westover to take the controls.

"About that time, the pilot walked out of the cockpit and went out on the wing and began to work with the motor," Westover later recalled in a speech on the "thrills of aviation" to the National Geographic Society.

"Well, you can imagine what happened to me," the general continued. "I got way up somewhere, the air began to climb through my helmet and I had the queerest sensation of dropping nowhere fast."

Westover said he used all his strength as a gymnast to control the plane, which listed due to the uneven weight distribution caused by the wing-walking pilot. After what seemed like an eternity to Westover, the pilot climbed back into the open cockpit and resumed control of the aircraft.

When the plane landed at Langley Field, the five-foot-five passenger angrily admonished the six-foot tall pilot, who had mistakenly inferred from their earlier conversation that Westover was an aviator.

"I don't see why you're cussing me out like this," the pilot protested. "I was only doing what every other pilot would do."

"Yes, you boob, but you didn't know that was the second time I had been in a plane and the first time I ever handled a control," Westover replied to the now-ashen faced flier.

Once he became a real pilot, Westover's career in the Army began to soar.

He served as commandant of Langley Field and the Air Corps Tactical School.

In 1930, he was promoted to lieutenant colonel, the temporary rank he had held a dozen years earlier.

By 1932, he was a brigadier general and assistant chief of the U.S. Army Air Corps. With a star on his collar, General Westover began pushing for modern all-metal bombers and the fighters to support them.

OSCAR WESTOVER considered himself well versed in aircraft construction. When Boeing introduced the low-wing B-9 all-metal monoplane, a new aerodynamic concept was introduced. Although the B-9 still had open cockpits, the twin-engine aircraft had the "big bomber" look which intrigued Westover and others who became advocates of strategic bombing.

At that time, advances in aviation were evolving through civilian channels, the Cleveland Air Races put the spotlight on speed and streamlined designs which blazed the trail for future fighter aircraft. The U.S. Post Office was flying the mail and developing a ground navigation network and the aircraft instruments for night and foul-weather flying. The airlines were using Fokker and Ford tri-motors and beginning to develop methods and routes for hauling passengers.

But the Depression-era defense budget was skinny and the Air Corps was underfunded and flying mostly aircraft which were obsolete by civilian standards. The Navy was in the Congressional limelight and working overtly and covertly to rein the Army's airpower.

THE QUESTION MARK referred to the number of hours the aircraft could remain aloft through aerial refueling. The Question Mark also symbolized the Army Air Corps' early efforts at garnering attention and influencing public opinion. Commanding the Fokker C-2 when the endurance record of nearly 151 hours was set in 1929, was Major Carl Spaatz. Also on board the Question Mark were Capt. Ira Eaker and Lt. Elwood Quesada, who later became head of the Air Force Reserve. The aircraft was refueled 43 times and 5,000 gallons of fuel were transferred. Two decades would pass before KC-97s were developed as aerial tankers. In the Fifties, Westover AFB became one of the most important tanker bases in the Strategic Air Command.

To understand what General Westover was advocating in those Depression years, one must pause to reassess the state of aviation only three decades after Kitty Hawk.

Air combat began in 1912 in Mexico when mercenaries flying for the rebels and federales took pot shots at each other.

Aviation advanced like dog-years during the World War, and by 1919 some 200,000 Americans had served in some segment of the air forces.

But flying and aircraft were beyond the realm of an American society still rooted in agriculture and just emerging as a world power. The "Roaring Twenties" brought barnstormers to Middle America, but the thought of air travel remained beyond the pale.

Air Mail boosted the concept, but the great leap forward is credited to Charles Lindberg, who captured the American imagination with his solo 33 1/2-hour trans-Atlantic flight in 1927.

He wasn't the first across the ocean, but he gained worldwide fame and demonstrated intercontinental travel would someday be possible.

The next major advance was a very quiet achievement. Lindberg's "Spirit of St. Louis" had used an efficient, air-cooled radial engine. A simple NACA cowling developed for that type of engine was the next step.

Unlike the Fokker or Ford tri-motor, the next generation aircraft -- like the famous GeeBee air racers built in Springfield -- used the cowling for greater airspeed and smoother air-flow over their wings.

Douglas Aircraft Co. entered into a contract with TWA in 1932 to produce a new type of cantilevered fuselage DC-1 with room for 12 passengers. A stretched version DC-2 came next and accommodated 14 passengers. Three years later, American Airlines ordered a 28-passenger version known as the DC-3. Douglas began working in 1935 on a bomber variant designated the B-18.

Few Americans had heard of General Westover until tragedy struck the Air Corps in early 1934 and his name was published in news accounts around the nation.

Senate investigators claimed the Post Office had favored the major airlines with air mail contracts. President Franklin D. Roosevelt stepped in on Feb. 9, 1934, and cancelled the air mail contracts for all 26 routes. He said the Army would begin carrying the mail 10 days later.

The air mail operation was given to Brig. Gen. Westover. He ordered ground crews to rapidly convert the Army's odd and obsolete collection of P-26 pursuit planes, B-9 bombers and observation craft for mail duty.

Few of the planes had reliable instruments and U.S. pilots trained to be deployed only in daylight with good flying conditions. Rain and gloom of night were not in the tactics handbook.

In the first seven days, six Air Corps pilots died.

OBSOLETE BOMBERS AND TRAINING became painfully apparent when General Westover was in charge of flying the mail. Critics were beginning to ask, "If the Air Corps can't fly at night or deliver the mail, how will it ever deliver bombs?"

Former World War ace Edward V. Rickenbacker criticized the president's decision as "legalized murder."

The focus of the national debate switched from contract irregularities to the inadequacies of the Air Corps.

General Westover stepped forward. "When you consider how the job was dumped in our laps, and how little warning we had," he said, "the men have done exceptionally well, particularly with our present equipment."

The predominantly anti-Roosevelt publishers ran columns of editorials calling for a return to civilian air mail flights instead of slaughtering the ill-equipped Air Corps.

On March 9, 1934, three crashes killed four more Army mail pilots. Another died on St. Patrick's Day.

President Roosevelt called in General Westover and his boss, Air Corps commander Maj. Gen. Benjamin Foulois. "General," the President said, "when are these airmail killings going to stop?"

"Only when the airplanes stop flying, Mr. President."

The generals endured a 10-minute presidential tirade on the inadequacies of the Air Corps. "The continuation of deaths in the Army Air Corps must stop," said the President.

Generals Foulois and Westover passed the word down to Lt. Col. Henry "Hap" Arnold, Maj. Carl Spaatz, Capt. Ira Eaker and Lt. Curtis LeMay -- and others who would later be linked to Westover AFB and earn fame in World War II -- that the Air Corps would further reduce mail routes to only nine and would only fly in good weather.

The deaths of 11 men in less than a month had aimed national and presidential attention on the shortcomings of America's military air force. The flying situation was unacceptable. If the Air Corps couldn't even deliver mail, how could Army pilots be expected to deliver bombs in wartime?

And the President, as we know from later histories, was keenly aware of the contemporary Nazi buildup and watchful of the emerging Luftwaffe.

Before the Army mail duty ended on May 7, 1934, the Air Corps took delivery of its first Martin B-10, an all-metal, enclosed-cockpit, 212-mph bomber capable of carrying 2,000 pounds. The B-10, which later was the first aircraft to land at Westover Field, was the beginning step in developing American airpower.

On Christmas Eve, 1935, Oscar "Tubby" Westover was appointed chief of the Air Corps and promoted to major general.

During the 1920s and early 1930s, Westover had studied everything he could about aviation in general and military flying in particular. He quickly became convinced that the airplane, like the crossbow and gunpowder before, would change warfare forever.

Now, with stars on his collar and rank to bolster his viewpoints, General Westover set out to put the theories he had developed into practice.

He quickly became a staunch and articulate national proponent of modern and well-equipped U.S. air forces. He advocated fleets of all-metal "flying fortress" bombers, and

NEW GENERATION B-10 bombers were developed by Martin. With enclosed cockpits, retractable landing gear and power-operated gun turrets, the B-10 had a top speed of 212 mph. A 10-bomber flight of B-10s, led by Lt. Col. "Hap" Arnold, flew from General Westover's headquarters in Washington to Fairbanks, Alaska. The 8,920-mile round-trip mission was designed to prove the Air Corps could reinforce outlying U.S. possessions.

fighters to support them. The general also called for more and better trained pilots and support personnel.

The Air Corps chief frequently told congressmen discussing aviation appropriations: "Develop a defense system in which the principal reliance is placed upon our own aircraft to destroy the enemy attacking forces."

Mindful of what he called "the disturbed international situation abroad and the violent efforts of most military powers to increase their military strengths" Westover warned that superpowers like Germany and Japan were outpacing the U.S. in the production of military aircraft.

"What we need is plenty of aircraft, and soon," he urged.

When General Westover assumed command of the Army Air Corps, the service had an inventory of 1,257 planes. He immediately proposed the number be increased to 2,320 by 1940, and urged continued aeronautical development.

"We cannot assume that the United States of America has a monopoly on the brains of the world," the general wrote in May 1938. "Today, our preeminent position in airplane design is due to the fact that for the past 10 years the United States government and private commercial airplane manufacturers have expended vast sums on airplane experimentation and development. The United States cannot always expect to occupy this enviable position without expenditure of money and effort."

The general soon became recognized for his work as an aviation expert. He was named a trustee of the National Geographic Society. In 1937, King Leopold III of Belgium named him a Commander of the Belgian Order and personally presented the decoration to Westover in Brussels.

Despite his administrative responsibilities overseeing the young Air Corps, Westover did not

MAJ. GEN. OSCAR WESTOVER
Chief of the Army Air Corps

want to become a "swivel chair" general. He loved the thrill of flying, and the cockpit -- any cockpit -- was the place he preferred to be.

Even as a two-star general, Tubby Westover flew every chance he got. He flew from city to city and base to base with his message.

Westover quickly found himself in the national spotlight as he called for the development of a large, long-range strategic bomber that was well-armed and could defend itself over enemy territory.

Like a politician on the stump, he criss-crossed the country, giving speeches about U.S. airpower. The message was consistent -- the U.S. needed new planes, ongoing aeronautical research and qualified aviators to fly the aircraft of the future.

Within a short time, Douglas was building wider and longer DC-3s which were flying regularly scheduled routes across the U.S. carrying passengers and mail.

The cantilevered construction, smooth nacelles and more powerful air-cooled radial engines were also being incorporated into large military aircraft, designed mostly as light and medium bombers.

General Westover wanted the trend to continue with even larger bombers with more payload capacity and greatly extended range. And he wanted next-generation fighters which could keep up with the big bombers.

Two months before he became the boss, an experimental Boeing Model 299 bomber was to be tested against a two-engined Douglas B-18, but the four-engined bomber crashed on its first Air Corps test flight.

***THE "MYSTERY PLANE"* Boeing's experimental Model 299 crashed on take-off Oct. 30, 1935 at Wright Field, Ohio.**

Boeing risked its own capital to develop a "flying fortress" type four-engine bomber such as General Westover had been advocating.

Model 299 first flew July 28, 1935. In late August, the 2,000-mile flight from Seattle to Dayton was made in nine hours at an average speed of 232 miles an hour.

The War Department wanted the Douglas B-18 *Bolo* -- terming it a "coastal defense bomber" to appease the Navy -- to become the mainstay of the Air Corps for the next five years and ordered 350 of the DC-3 variants. But General Westover argued forcefully for the four-engined Boeing Model 299, known to the Air Corps as the Y1B-17.

In early 1937, the War Department relented and allowed the Air Corps to contract with Boeing for 13 of the B-17 "flying fortresses." The first B-17 flew out from Seattle and landed at Langley Field on March 1, 1937.

With a top speed of 230 mph and a service ceiling of 25,000 feet, the 2,200-mile range of the Boeing B-17 answered the first part of General Westover's dream of a strategic bomber. Further modifications increased speed, ceiling and bombload to 17,600 pounds with 11 machineguns of .50-caliber.

But the role of Westover's Air Corps was always limited by a fight for institutional survival in the shadow of the powerful influence of the U.S. Navy. While Westover was commander of the Air Corps, the aviation battle between the services became so intense President Roosevelt ordered the two branches to square off in an operational test. Joint Air Exercise No. 4 was held in August 1937 and pitted B-10s, B-18s and one of the first B-17s, against a battleship somewhere in 100,000 square miles of ocean. The Navy stacked the deck by selecting a foggy area off the West Coast and requiring the bombers to fly above a thousand feet.

A B-17 with Lt. Curtis E. LeMay aboard as navigator flew below the imposed ceiling, found the USS Utah and hit the battleship with water balloons. Navy officers screamed foul, saying the Air Corps had conducted a "sneak attack." The Navy moved to classify the test, but unnamed Air Corps officers leaked the results and photographs to the news media.

When another test was held the following year, the Italian liner Rex was found 600 miles out in the Atlantic by three B-17s.

REX INTERCEPTED, May, 1938

Aboard the three B-17 bombers were reporters from NBC radio, the *New York Times*, the *Washington Post* and the *New York Herald-Tribune.*

Westover also talked secretly of a "hemispheric defense bomber" a super long-range, high-altitude, multi-engined craft powerful enough to span continents. But General Westover never lived long enough to see the B-29 *Superfortress* or the B-36 *Peacemaker.*

He also argued successfully in 1937 for buying the DeSeversky P-35 and Curtis P-36, the first all-metal, enclosed-cockpit fighters with retractable landing gear. Powerplants and armament were inadequate but the 210-plane order for P-36 *Mohawks* was the largest Air Corps fighter purchase since the World War.

THE CURTIS P-36 MOHAWK was a major purchase by the Army Air Corps. The 210-plane buy started the service on an elaborate modern training program that eventually created the leaders of the pilots who would fly the 13,785 P-40 Warhawks; 15,683 P-47 Thunderbolts; and 13,653 P-51 Mustangs.

First modern fighters

DeSEVERSKY P-35s had the classic lines, like those of the P-36, which were to be found during World War II on variations of many Navy and Air Forces aircraft. The first production models reached the Army Air Corps in July 1937 and were 50 mph faster than the P-26.

In the late summer of 1938, Lockheed Aircraft Corp. was developing an advanced long-range, twin-engined fighter which could escort the new B-17s over enemy territory. General Westover was keenly interested in the radical twin-boom fighter known then as Lockheed Model 22. He planned to visit the Lockheed plant in the autumn to see the fighter designed by Clarence L. "Kelly" Johnson.

General Westover also toured Army Air Corps installations throughout the country, talking shop with the military aviators at each post.

He frequently flew his own plane, a Northrop A-17, accompanied by his long-time crew chief and friend, TSgt. Samuel Hymes. Westover's plane was one of just two A-17s in the Air Corp's inventory.

On a visit to Lowry Field near Denver on Sept. 10, 1938, the Air Corps chief and Sergeant Hymes survived a brush with disaster.

General Westover had just inspected the Air Corps technical school at Lowry and had taken off in his A-17 from the Denver Municipal Airport when the plane developed a serious oil leak.

A front page news story that day in the *Denver Post* noted: "Handling his specially built, attack-type aircraft with all the skill fitting his position as chief of the United States Army Air Corps, Maj. Gen. Oscar Westover of Washington, D.C., brought his ship down safely at municipal airport Saturday morning after it developed a serious oil leak while in midair and was in danger of catching fire."

The article also reported the closed-in cockpit of the A-17 was completely covered by the oil spray, obscuring Westover's vision.

"I threw back the covering and almost stood up in the plane to see to turn it back to the airport," said General Westover. "I was afraid

Trouble, and Plenty of It caused oil to spurt from the motor of Maj. Gen. Oscar Westover's plane Saturday morning. General Westover had a narrow escape in bringing the plane safely back to the airport without it catching fire. Here is shown a group of the ground crew working on the plane trying to locate the trouble.

at any minute that it might burst into flame."

Westover, who was covered with oil when he landed, lost his flying helmet and radio earphones when he threw open the glassed-in covering over his cockpit.

Mechanics later said the mishap was caused either by a break in the plane's oil line or a loose connection in the motor's fuel system.

A balloonist and airplane pilot, Oscar Westover knew danger was a constant companion in the air. That fact, he said, drew him closer to God.

"I have stated repeatedly," he said in one speech, "that I know of no group of men or women who commune more sincerely with God than those who fly in the air and are entrusted with the responsibility of the lives of others whom they are flying. This communion is instinctive in many ways and many cases. At other times it is produced by harrowing circumstances and realization that after all human skill, in fact all our traits are not infallible, that all our reliance must be upon that support and guidance and that re- establishment of confidence which aviators find in communion with God."

WESTOVER AND TSGT. SAMUEL HYMES

Just 11 days after the near disaster over Denver, General Westover and Sergeant Hymes were flying in the A-17 toward the Lockheed plant near Burbank, Calif.

The Air Corps commander had been feted the evening before at the "Men with Wings" aviation ball in Los Angeles. He had just left March Field, where he presented the "Daedalian Award" for flying safety to the 19th Bombardment Group.

As General Westover's aircraft approached the Lockheed Corporation Airport, the Northrop attack fighter suddenly experienced difficulty.

"The ship was, I imagine, about 300 feet high and a half-mile northwest of the field when it went into a hairpin turn," said an eyewitness in the Lockheed control tower.

"Suddenly, I saw the entire ship quiver," he said. "The ship nosed downward in the start of a spin. You could see the pilot was fighting desperately to regain control...It all happened so quickly -- there were just a few seconds between the time the ship started to skip off and the time it hit the ground."

Witnesses said the A-17 hit with the "roar of a bursting bomb," in the front yard of a home half a mile from the Lockheed plant.

General Westover and Sergeant Hymes were killed instantly in the fiery crash on Sept. 21, 1938.

Smashed Plane in Which Air Chief and Pilot Perished

Associated Press WIREPHOTO

Wreckage of the Army attack plane in which Maj. Gen. Oscar Westover and his pilot, Sergt. Sam Hynes, were killed yesterday when they crashed into a bungalow court at Burbank, Calif., while attempting to land at flying field of the Lockheed plane factory. The ship was still smoldering when photo was made

The following day, General "Hap" Arnold was named chief of the Army Air Corps.

According to Patricia Freudenthal, the Westover family received hundreds of condolence letters and tributes after her father died. "People just loved him," she said. "He was a fun person with a great sense of humor."

The 55-year-old Army Air Corps commander was one of only two general officers to die in the line of duty since the Civil

Three generations of Westovers

His pilot grandfather helped found the fledgling Army Air Corps. His pilot father helped run two major Air Force commands.

From them, Col. Timothy O. Westover inherited a love of flying, an Air Force name and a large chunk of Air Force history.

For nearly 30 years, the colonel has been living the legacy and helping to write some Air Force history of his own.

Colonel Westover, a veteran pilot and Distinguished Flying Cross winner, is stationed with the Air Force Systems Command at Wright-Patterson AFB, Ohio.

He is the program director for the LANTIRN (Low Altitude Navigation and Targeting Infrared System for Night)) System Program Office at Wright-Patterson.

The 49-year-old officer is the grandson of Maj. Gen. Oscar Westover and the son of Lt. Gen. Charles Bainbridge Westover.

Colonel Westover's father, Charles Bainbridge Westover, was an Air Force three-star general who served as vice commander of both the Tactical Air Command and the Air Defense Command.

Following in the footsteps of his famous ancestors was a matter of "tradition," Colonel Westover said. "The Air Force is a way of life in my family," Colonel Westover said. "There's a lot of tradition..."

Despite the generation gap, the careers of all three Westover men have been strikingly similar.

Colonel Westover graduated from the Air Force Academy in 1964. Oscar Westover was a member of the class of 1906 at the U.S. Military Academy, and Charles Westover graduated from West Point in 1937.

Each Westover won his wings as a military pilot. Each worked with strategic bombers. Each held an important position and each helped to advance modern military airpower.

Colonel Westover is a test pilot who has flown more than 3600

Colonel Timothy Oscar Westover

War. His death also created a huge void for the Army and the nation.

"In my opinion, the Air Corps owed more to General Westover than any other officer who has ever served in the Corps," said Maj. Gen. George Van Horn Mosely, commander of the Army's Fourth Corps Area.

Lieutenant General Charles Bainbridge Westover

hours in 26 types of aircraft during his career. He logged 423 combat missions in Vietnam, where he earned the DFC and Bronze Star.

His present assignment with the System Command's LANTIRN program -- designed to enhance the night-fightingcapability of F-I5E and F-I6 fighter jets -- puts him on the cutting edge of technology.

According to the colonel, the LANTIRN system produced by the Martin Marietta Corp. consists of two separate equipment pods mounted under the fighter's fuselage.

The navigation pod contains terrain following radar and infrared sensors which enhance the jets' capability to fly at low altitudes in the dark. The targeting pod contains sensitive, high-tech equipment which allows the pilot to detect and track targets at night.

Colonel Westover's involvement with the LANTIRN program would undoubtedly make Oscar Westover smile.

More than 50 years earlier his grandfather advocated the development of a revolutionary type of aircraft -- the "flying fortress type of bomber" for long-distance missions.

Although Colonel Westover never met his grandfather -- Oscar died two years before he was born -- he learned about him from his father.

"Both my father and grandfather loved the military," the colonel said. "Loyalty and honesty meant a great deal to both of them."

The colonel said that he also studied his grandfather "a little bit" while he was a cadet at the Air Force Academy. "There was more emphasis placed on airmen like Billy Mitchell and Hap Arnold," he said.

General Arnold, a former assistant to Oscar, suggested the Northeast Air Base be named in honor of Westover.

Colonel Westover's father, Charles, was the acting TAC commander during the Cuban missile crisis in l962. He retired from the Air Force in l967 as vice commander of ADC.

Following his father's death in 1978, the colonel inherited Oscar Westover's papers and military memorabilia. He subsequently donated the collection to the Air Force Historical Research Center at Maxwell AFB, Ala.

Scores of military officers and foreign dignitaries attended General Westover's funeral at Arlington National Cemetery on Sept. 27, 1938. Fifty military bombers and pursuit planes circled overhead as the legendary aviator was laid to rest with full military honors.

Mrs. Freudenthal said her father's burial site was selected with considerable thought. "My mother picked it because it was on a slope that overlooked his old office in the Munitions building, and because airplanes regularly flew over the spot," she said.

America's "Number One Flier" never lived to see the air base named in his honor -- nor the large bombers and cargo planes which have flown from there on thousands of missions for half a century.

But the Chicopee air base and its aircraft are perhaps the most enduring monuments to "Tubby" Westover's vision and courage.

Four months after his death, the first Lockheed Model 22 -- redesignated the P-38 *Lightning* -- made its maiden flight.

Eleven months after that, Westover's longtime pal, General "Hap" Arnold, announced on Dec. 1, 1939, that the airfield at the modern super-airbase under construction in Massachusetts would be named in honor of the visionary pilot who had guided America's concept of long-range airpower.

The base was aptly named in those dark days before Pearl Harbor when the U.S. War Department worried that England would not survive the Nazi *blitzkreig* and President Roosevelt was trying to spur America's aviation industry into expansion.

A half century later one sometimes forgets that Westover Field was being built at the height of the "Battle of Britain" when Spitfires dueled with Messerschmits and London was on fire.

The Springfield *Daily News* noted at the time of the base dedication in April 1940 that Westover Field could be used for B-17s to bomb Greenland. One of the first fighter groups to be assigned to the base flew the P-38 *Lightnings* which had fascinated General Oscar Westover. And the first aircraft officially assigned to the base was a Northrop A-17.

FACING PAGE -- Oscar Westover, in civilian clothes, speaks with two Vultee Aircraft Co. executives at Downey Field, Calif., a few hours before he and Sam Hymes left for the Lockheed plant at Burbank. This last photograph of General Westover was sent to his widow on Oct. 14, 1938, by Maj. Gen. "Hap" Arnold, who identified the man in shirtsleeves as Mr. Millar and the other Vultee executive as John I. Wagner.

(Freudenthal Collection)

APRIL 6, 1940, the day Westover Field was dedicated, Staff Sergeant F. E. "Ed" Maliar was a young "aerial engineer" with the 99th Bomb Squadron stationed at Mitchel Field, N.Y. On the morning of April 6 he and fellow crew members preflighted their Douglas B-18A Bolo for the flight to Massachusetts. Five other Bolos from the 99th were joined by eight other B-18As from Mitchel and they formed aloft with planes from Selfridge Field, Mich., for the trip to what would become Westover Field. Four B-18As are visible in this photograph.

"Note the several long drying barns for the tobacco, the major crop in the area," said Maliar. "The roadways are lined with cars of people who came out for the dedication."

The photograph taken from the B-18A shows a view toward the southwest. Ludlow City Road is at the lower left, the roadway running to the upper right is Prospect Street, which intersects with Granby Road toward Pendleton Avenue.

PREPARING WESTOVER FOR WAR

European conflict spurs growth at the 'super-airbase'

Just before Westover Field was dedicated in April 1940, some 1,400 workers of the Civilian Conservation Corps and Works Projects Administration arrived in Chicopee to begin a $750,000 project clearing tobacco and farm fields for the $3.6 million Northeast Air Base, announced seven months before by Secretary of War Harry H. Woodring.

On April 6, 1940 -- which was designated "Army Day" -- many Air Corps officers, state and city dignitaries and thousands of Greater Springfield residents gathered for the official dedication ceremonies. Eighteen silver B-18 bombers flew overhead as bandsmen played, and the base was launched on an historic course which would intersect often with modern U.S. history.

In keeping with the American military tradition of naming ships, tanks, forts and airfields to honor departed heroes, the flying field at the Northeast Air Base approved by the Wilcox Act was officially named in honor of Arnold's pal and long-time boss, Maj. Gen. Oscar Westover, the former Army private who had learned to fly.

Although only the airstrip was designated Westover Field, the entire Northeast Air Base was known locally and eventually within the Air Corps, and later the Army Air Forces, as Westover Field. Shortly after the U.S. Air Force was formed in 1947, the entire installation was officially designated Westover Air Force Base. But even in 1990, Massachusetts Turnpike signs designate the highway exit for "Westover Field."

B-18A BOLOS wing southward over Springfield after buzzing the tobacco plains between Ludlow and Chicopee. The bombers flew to the dedication ceremony from Mitchel Field, N.Y., and Selfridge Field, Mich.

Dedication day

Neil O'Leary was a 12-year-old Boy Scout when Westover Field was dedicated and remembers "some general strutting around. Army Air Corps generals wore puttees and bloused knickers. I remember the speeches and the most exiciting flyover -- a string of 18 B-18s. This was one of America's first metal bombers. They were silver with the Army Air Corps insignia on them and they flew over for quite a long time and this was very exciting.

"In the middle of a speech by a general, the group broke up and there was a lot of excitement. We were 15 to 20 feet from the podium and we realized a skunk had come out of the high grass. He was shaken up and he opened up that crowd. Broke up the whole ceremony. So we as scouts quickly realized some heroics were called for and we were trying to muster up enough courage to go shoo the skunk away. But we didn't have to; the skunk went away. The general made quite a joke out of that. I think he said something like, 'Not everybody here is ready to accept us yet.'

Following the groundbreaking for the photo lab, there were build ings popping up all over....there were orange tags on trees and zig-zagging the area for miles."

A few weeks after the dedication, as the war spread in Europe and China, the Springfield *Daily News* reported on May 25, 1940, "With the war situation more tense abroad and now brought to a climax in this country, it appears likely that Congress will now earmark in its defense appropriation a large sum -- possibly $5 million -- as an added appropriation for the Northeast air base construction and maintenance."

Owners of the seven square miles surrounding the intersection of Ludlow City Road and Prospect Street -- today the intersection of Runways 23 and 33 -- were paid slightly more than $2 million after their properties were taken by condemnation.

Tents were erected and abandoned farmhouses were used as offices during the early phase of construction. Maj. Murdock A. McFadden was "constructing quartermaster" for the project and worked out of an office at the former Chicopee Isolation Hospital in Aldenville.

At the time, Westover was expected to become the major aircraft overhaul point on the East Coast. The German army and air force were rapidly conquering Norway, Holland, Belgium and France, and the Army Air Corps was committed to a plan calling for 54 combat groups. The American plan was no sooner underway when a revised two-year plan was unveiled that called for 84 combat groups equipped with 7,800 aircraft manned and maintained by 400,000 men by June 1942.

Col. Richard H. Ballard was appointed the first commander of Westover Field, and one of his first projects was to erect a large flagpole off Sheridan Street to mark the entrance to what would become one of the most important Army Air Corps fields.

The tempo of air base construction increased even more after German troops smashed into Holland, Belgium and Luxembourg at dawn on May 10, 1940. President Roosevelt delivered his national defense message soon after and at Westover the first runway was carved in May by giant graders.

In early July, a New Jersey firm was awarded a $1.6 million contract for erecting the huge steel and brick hangars which are today Hangars 1, 3, 5, 7, and 9. There were plans then to build five more hangars on what is today Westover's North Ramp but the fortunes of war made them unnecessary.

By the time France fell in June 1940, military aircraft production in the U.S. was slightly more than 400 aircraft a month. By the end of the following year the figure was up to nearly 2,500 a month and was steadily rising. Westover Field was going to become an important jump-off point for thousands of aircraft headed for Europe.

Five enlisted men arrived from Maxwell Field, Ala., on July 29, 1940, and were the first Air Corps men to be stationed at Westover Field. Among them were Glen Flanders of Mississippi and William Kilpatrick of Alabama who arrived at the base in a 1936 Plymouth. Both men became aircraft mechanics, served in England during the war and eventually retired from the Air Force.

The first permanent building officially accepted by the War Department was the brick photographic laboratory in October 1940. On Oct. 10, a B-18 *Bolo* became the first bomber to land at Westover Field. Aboard the aircraft was U.S. Rep. Charles R. Clason of Springfield who had aided Mayor Stonina in obtaining the War Department's approval for constructing the base.

In early December, base commander Colonel Ballard landed an A-17 attack ship at Westover and the aircraft, the same as General Westover had been flying when he and Sergeant Hymes were killed, was the first plane officially assigned to the base.

A week before Christmas 1940, Brig. Gen. John B. Brooks arrived at Westover with the assignment of running the 4th Bombardment Wing which was coming to the base within six months. General Brooks had been a major in operations on the War Department General Staff in 1935 when Oscar Westover was commanding General Headquarters Air Force.

By the end of 1940, total Army Air Corps strength at Westover Field was 68 officers and 1,124 enlisted assigned to two ordnance units, the 26th Air Base Group, a finance office, a medical department, one quartermaster regiment with an attached battalion, and an aviation signal corps service company.

In January 1941, more than 1,600 construction workers employed by 10 firms put in 230,000 hours at

From left: Glen E. Flanders of Mississiippi, C.H. Huggins of Florida, William J. Maness of Tennessee, and Bill Kilpatrick of Alabama.

Nearly 50 years after he stepped off an Army truck at Westover Field in the summer of 1940 as an 18-year-old recruit, retired Master Sergeant Glen E. Flanders returned to Westover AFB.

"There were no runways, no hangars here. I remember a pilot landing and taking off from a tobacco field," he said. "There were only two wooden barracks, a mess hall and a supply building." When Flanders arrived at Westover with William Kilpatrick their first duty, as Army Air Corps cooks, was to unpack the pots and pans to feed approximately 50 more men coming to the new base from Alabama.

"The first plane to land at Westover was by mistake," said Kilpatrick. "When it landed, we all went out to it and the pilot said he was looking for Barnes Airport. One of the first aircraft assigned to Westover was flown in by a Lieutenant Golf and I believe it was an 0-38. It was a biplane with two open cockpits...later we got a B-18 and that was a big deal."

Flanders, a Mississippi native, and Kilpatrick from Alabama, became aircraft mechanics and were stationed at Westover until 1942, when they were transferred to RAF Alconbury in England with the 36th Material Squadron. Their squadron was later stationed at Paddington, England, and eventually in France. After the war ended, Flanders returned to the base in 1945 and remained at Westover AFB until 1957. While at Westover Field he met and married Alexandra Gayda of Holyoke.

Westover as WPA workers continued pulling stumps and filling cranberry bogs near Granby for another 144,000 hours. The Corps of Engineers worked 13,800 hours that month and other contractors put in 72,575 hours, according to a report by the Providence Engineer District.

As the "Battle of Britain" raged, Westover Field became touted as "America's biggest air base." The *Boston Sunday Advertiser* noted on March 23, 1941, that the largest hangar at Mitchell Field would fit inside one of Westover's five hangars and the main landing strip was twice the length of the runway at Boston Airport.

Westover Field was designed to have four runways, two more than two miles long and two about a mile and a half long. At the time, LaGuardia Field in New York was considered the nation's most modern airport with a 6,000-foot runway.

Within a year, five of the nine proposed hangars were completed, along with a brick fire station and 106 buildings housing nearly 3,000 troops, a 500,000-gallon watertower complete with rotating beacon, and 12 miles of roads for the hundreds of tank trucks transporting aviation gasoline to storage containers for 200,000 gallons of fuel and 10,000 gallons of oil.

Capt. Frank D. Huyler, Westover's public relations officer, predicted in March 1941 that about 100 bombers would soon arrive at the base. He said B-17s would be able to fly non-stop to Russia from the Chicopee air base.

A *Boston Advertiser* reporter noted at the time, "In the event that enemy aircraft carriers maneuvered off the New England coast, fleets of these heavy bombers from Chicopee would streak out to sea and drop their high explosive cargoes. In the same way, possible enemy bases in an invaded Canada

FIRST BOMBER LANDS at Westover Field. Congressman Charles R. Clason arrived Oct. 10, 1940, aboard a Martin B-18 bomber.

First B-17 arrives at Westover Field, June 10, 1941.

or as far east as Greenland could be blasted by the American craft."

A 100-vehicle convoy made a 95-mile night trip from Westover and around the Quabbin Reservoir under blackout conditions in May of 1941. Western Massachusetts residents were required to live under blackout conditions during Civil Defense exercises and became increasingly familiar with the "ghostly blue lights" of brown military vehicles as the populace trained for air attack conditions.

In early June, Westover was one of two dozen new bases officially approved for occupancy by the Army Air Corps. On June 10, 1941, five silver B-17s with red-and-white rudders flew low over Springfield and came to roost at Westover Field as the first assigned aircraft of the 4th Bombardment Wing. Traffic stopped along highways as people craned their necks to view the streamlined bombers that Maj. Gen. Oscar Westover had advocated.

One of the copilots on that first B-17 flight to Westover was a pudgy major identified in the Springfield *Union* as C.E. LeMay. He soon took command of the 7th Bombardment Squadron at the field and later, as a lieutenant colonel, was operations officer of the Eighth Air Force at Westover before shipping out to England. He later became famous as cigar-chomping General Curtis Emerson LeMay, commander of the Strategic Air Command, and would have a profound effect on Westover and America during the Cold War.

The arrival of the first B-17s marked the true activation of the base. Brig. Gen. John B. Brooks was the 4th Bombardment Wing commander and in charge of the 34th Bombardment Group, 26th Air Base Group -- which coincided with the 26th "Yankee" Division of the Massachusetts National Guard -- and 1st Reconnaissance Squadron, plus other smaller units.

London was being hammered in June 1941 by German bombers and hundreds of Greater Springfield residents listened to their radios for news reports.

Springfield was attacked by paper bombs on June 16, 1941, to raise money for the USO. The bomb-shaped leaflets dropped from a formation of B-17s read: "No, this isn't a bomb. It was dropped not to destroy, but to build. Dropped by one of your defenders from one of the planes that guard our nation's skyways."

On June 20, 1941, the Army Air Corps and GHQ Air Force were combined into the Army Air Forces. By the end of the month, the 13-mile fence around Westover Field was complete and the base was officially closed to the public. On July 7, American troops joined British troops in the occupation of Iceland, supposedly to protect airfields there so American convoys could have air cover. Most Americans knew at this point that the United States would soon be engaged in the war in Europe. The German consulate in Boston had been closed June 16 and President Roosevelt issued an order July 25 freezing Japanese assets and halting all trade with Japan. The fence around Westover Field became a local symbol that the war was coming to Western Massachusetts.

At the end of July 1941, Maj. Gen. Henry "Hap Arnold, chief of the new AAF, inspected the base and watched practice bombing runs made at altitudes of 300 and 10,000 feet. He told reporters of the War Department decision to make Westover "the strongest base in the East" and a ferry point for aircraft bound for England.

A balloon launch by the 4th Bomb Wing weather section set a world altitude record on Sept. 5, 1941. The hydrogen weather balloon was observed through a theodolite while ascending to 41.31 miles. The previous record had been 24.18 miles.

For a short while, the all-black 369th Antiaircraft Coast Artillery Regiment stationed at Westover. During the racially segregated era of World War II, the 1,624 officers and men of the 369th were known as "Harlem's Finest." Three years later there was a "colored" recreation hall on base.

FIRST B-24 LIBERATORS arrived at Westover in November 1941 for anti-submarine missions to protect convoys carrying Lend-Lease supplies to Britain.

By October 1941 the Germans were within 60 miles of Moscow and there were 35 B-17s on the ramp at Westover. On Oct. 31, the *USS Ruben James*, a destroyer on convoy duty was torpedoed by a German U-boat and 100 American lives were lost in the North Atlantic.

In November, the first B-24 *Liberators* began arriving at Westover and were assigned to a heavy bombardment group at the base. The B-24s were to be used for anti-submarine warfare right off the coast of New England and Canada.

The first overseas unit deployment from Westover took place in November 1941 when the 803rd Engineering Battalion of the 4th Bomb Wing departed the base.

Soon after two tons of turkey were consumed by Army troops for Thanksgiving 1941, anti-aircraft guns were installed to protect the base and machinegun positions were set up along roads leading into the base.

By the next week, the United States was at war.

ST. PAUL'S CATHEDRAL stands amid the flames and smoke of burning London on the night of Dec. 29,1940. Many Americans believed England would not withstand the Nazi onslaught and the war would soon be fought from bases such as Westover Field.

WESTOVER DURING THE WAR YEARS

Thousands of young airmen pass through the base on their way to Europe and 'the numbers game' over Germany

The day after Pearl Harbor was attacked, there were lines of volunteers outside recruiting offices throughout New England. Truckloads of recruits from Greater Springfield communities surrounding Westover began arriving at the base.

Security was tightened, local newspaper reporters were ordered to return their press cards and all base personnel were photographed. More aircraft arrived at the base and reports of Westover men being killed in the Pacific and European theaters soon became standard fare in local newspapers.

Westover Field became an assembly point for Curtis P-40 fighters. Retired MSgt. Glen E. Flanders remembers the aircraft arriving on flatbed railroad cars with the fuselage in one box and the wings and propeller in another. "After they were assembled, inspected and tested, they were flown away. The people who worked on this project did not work by the hour or day, but to the limit of their endurance," he said.

The base was the training ground for many fighter units during the early stage of the war.

As progress was made on Westover Field in early 1942, the base became important for anti-submarine warfare. Westover launched a minimum of four armed bomber patrols a day, from dawn until dark. Although pilots who were assigned to the base at the time remember many depth charges being dropped by Westover planes, no U-boat kills were reported.

Ben Mathis was an Army Air Forces captain assigned to anti-submarine operations in B-24s stationed at Westover. "We would do a lot of touch-and-go landings and a lot of instrument flying, night flying, to get more used to the airplane. Eventually these missions became quite lengthy. Some of them were 15 hours, so there was a lot of night flying, a lot of weather flying, a lot of instrument flying....It was kind of a boring job because we usually carried a crew of nine and our sole job was to watch the surface of the water to see if there was any indication of 'white silk' or a periscope. Almost every day in the summer of 1942 and '43 there was a lot of U-boat activity in the North Atlantic and they sunk a lot of ships. We flew up around Nova Scotia and back to Westover. Some went as far south as Bermuda and some went directly out to beyond Halifax."

To practice depth charge runs, Westover crews would train over Quabbin Reservoir. A bright yellow target was erected in the reservoir and the B-24s would make low-level runs from various directions and altitudes with dummy depth charges to train both the bombardiers and the crews. Westover Field had a four-boat navy assigned to the Quabbin for aircrew rescue operations. The boats were assigned to the bomb and gunnery range detachment which was activated in July 1942. The range included a photo gunnery target which was built for the fighter pilots, a submarine conning tower for the B-24 and other

***CONSTRUCTION** was still underway when the Japanese attacked Pearl Harbor on Dec. 7, 1941.*

anti-submarine bombers, and targets of various types, such as high altitude bulls-eyes and camera targets. Two tow targets were constructed at the Quabbin with paddlewheels to simulate the wake of a submarine periscope. The bomb and gunnery range was a 24-hour operation.

By September 1943 the U.S. Navy had ample crews for the anti- submarine mission, and Westover's B-24 crews were quickly trained for high-altitude bombing.

The tide of war was turning against the Axis powers and the War Department became more confident the base would not be needed for coastal defense. The field became an advanced bomber training base and Westover's main product was air crews.

As graduates from various air combat training courses arrived at Westover, the men were teamed to provide the bomber crews of B-17 *Flying Fortresses* and B-24 *Liberators* assigned to the Eighth Air Force in England.

Their time at Westover allowed the pilots, copilots, bombardiers, navigators, flight engineers and gunners to fly as

QUABBIN RESERVOIR was used for high- and low-altitude bombing practice by B-17s and for anti-submarine warfare training by B-24s.

a team and begin building the spirit needed to survive in the skies over Europe.

As thousands of aircrews and aircraft departed Westover for assignments in North Africa, England, and the Middle East, one of the final touches at Westover was the installation of a brand-new Norden bombsight.

The Norden Bombsight

The Norden bombsight was the American key to strategic bombing. Prior to the invention of the Norden and Sperry bombsights, accurate high-altitude bombing was considered impossible by military strategists. In the early 1930s, simple bomb-aiming sights guaranteed only fair accuracy in hitting targets from altitudes below 5,000 feet. But at the heights thought then to be safe from anti-aircraft, bombing accuracy was nearly impossible.

Calculating the release point was a formidable task while a bomber traveled rapidly in three dimensions and rotated about three axes while being buffeted by air turbulence -- especially the turbulence caused by formation flying.

According to one bombardier manual, "The path of the dropped bomb was a function of gravity acceleration and aircraft speed, modified by altitude, wind direction and the ballistics of the bomb. The bombadier's problem was not simply an airborne version of a gunner hitting a moving target; bombing was the equivalent of aiming a moving gun to fire a variable powder charge from a platform evading groundfire and enemy fighters."

Prior to the Norden bombsight, the usual method was to use a Pilot Direction Indicator, which allowed the bombardier to track the target and press buttons that moved a needle on the pilot's panel and told him to fly left or right.

In Europe during the early days of World War II, the British had attempted accurate daylight raids in early 1941, but severe losses caused the RAF to abandon daylight formations and resort to night bombing by streams of aircraft which destroyed areas rather than specific targets.

The American plan under General Arnold was to manufacture bombers faster than the Luftwaffe could knock them down and to use massive formations of bombers in daylight. With the Norden bombsight installed at Westover Field -- and ideal conditions -- an Eighth Air Force B-17 could drop a bomb from 30,000 feet and hit a target.

The Norden bombsight, which did so much to win the war, was also symbolic of the Army Air Corps long-standing battle with the U.S. Navy.

During World War I, Carl L. Norden was a Navy consultant developing ship gyroscopes. Following the war he entered into a partnership with a former Army colonel, Theodore H. Barth. For the next few years, Norden designed bombsights and Barth tested them. In 1928 the U.S. Navy ordered 40 of the

NORDEN BOMBSIGHT

A tunnel at Westover between the current communications building and Hangar 1 was reportedly used to move the highly classified Norden bombsights -- actually a mechanical computer -- to the B-17 and B-24 bombers. A brand new Norden bombsight was installed on each bomber before the aircraft left Westover Field for the European Theater.

top secret bombsights and the two men formed Carl L. Norden Inc., a non-descriptive company name which would not bring attention to their secret manufacturing enterprise.

Their first Norden bombsight was delivered to the Navy in 1931, but there were problems with maintenance and the service endurance of the device. Their contract, however, specified that the Navy would purchase sights only from Norden, and Norden could only sell to the Navy. In effect, Norden and Barth had made their firm a manufacturing branch of the Navy, and the Air Corps was denied access to the bombsights.

By the time Oscar Westover had become chief of the Air Corps, the Navy had installed Norden bombsights on the slow-moving medium-altitude PBY bombers, and the Air Corps had difficulty developing a bombsight for high-altitude bombers such as the Boeing 299, which became the B-17.

In the summer of 1937, a secret meeting was held in New York City by Navy, Air Corps and Norden executives. Norden offered to set up an Air Corps manufacturing operation, but the Navy did not accept the offer. The problem endured until the summer of 1943 when Norden had mass-produced enough Navy bombsights and had enough of a surplus for the Army Air Forces to standardize on the Norden M-series bombsight.

In 1940 and 1941 the early Norden Mark XV bombsights were finally installed on a few B-17s. Some of the B-17s delivered in the late 1930s were equipped with a Sperry autopilot which could be linked to a bombsight. The Norden company and Honeywell also developed autopilot systems for use with the Norden bombsight.

Once the Army Air Forces settled on the Norden M-series bombsight in the summer of 1943, many of the sergeant bombardiers were warranted to flight officer and the service developed a "top secret" legend about the device.

The tunnel at Westover Field helped nurture that mystique.

With so many aircrews passing through Westover and facing the horrors of the air war over Germany, something was needed here to boost their spirits and to assure the crews would give their all to get their bomber over the target. Just as Hitler would later have "secret weapons" to bolster the faith of Germans, the Army Air Forces relied on the "secret weapon" of the Norden bombsight.

Bombardiers had to swear an oath to "protect the secrecy of the American bombsight, if need be with my life itself."

The top portion of the Norden, dubbed "the football" by aircrews, was removed from the bottom stabilizer whenever the aircraft was on the ground, and was escorted by armed guards to a secure facility on the base.

Before the war was over, Norden M-series bombsights on three-quarters of all Army Air Forces bombers had been downgraded from top secret to only confidential, but the secrecy legend lived on.

What was classified at the end of the war were reports that high altitude bombing with the Norden was not really that effective due to the poor weather over Central Europe, and American engineers were already hard at work on secret weapons projects such as radar-guided and television-guided bombs.

And the tunnel at Westover Field was part of the cultivated mystique of the legendary Norden device which bombardiers were sworn to protect.

Stonina continued his support for Westover Field

After the war began, Mayor Anthony J. Stonina attempted to join the Army. Although he was well known in Washington because of his efforts in siting Westover Field in Chicopee, the military wouldn't accept him.

"They told him he was more important out trying to sell bonds," said his son-in-law, Ben Mathis of Chicopee. "He'd go to Chicago. He'd go to Pittsburgh. He'd go anywhere there was a predominance of people of Polish origin; and he was a very effective speaker. He could really motivate people in his own ethnic group to buy war bonds."

Stonina would tell of how he came to America from Poland at age 12 and how he used to live in a thatched roof cottage in Poland. He was proud to be a self-educated man, the youngest mayor of Polish origin and a three-term mayor of Chicopee.

"He helped out a lot of people," said Mathis. "Particularly people who came from the old country and couldn't speak the language. He would give them a job as a janitor or something until they could get going."

Stonina always had a special place in his heart for the military. "I guess in Europe," said Mathis, "a military man was about as distinguished as you could get and (Stonina) always thought a great deal of military men." One of the early base commanders at Westover was Colonel Joe Barzsinski, whose father was an Army general. "Stonina thought that was great."

Mathis recalled that Stonina "was a proud man. He was proud of Westover. He wouldn't let people forget that either.

"In the course of a conversation, you couldn't have a conversation with him for more than five minutes -- even a stranger -- that something would come up and he would say, 'By the way, that reminds me of the time when we were negotiating for the Westover Field...' and he would tell the stranger how he managed to get Westover field for Chicopee."

One aspect of the Stonina story was how he managed to work with the Polish farmers who owned the fertile fields which were to become Westover Field.

Mathis said the Polish farmers "all felt he was instrumental in the government taking their land....There was severe criticism of him....He made enough enemies to defeat him after his first term, and all that was based on opposition that people had for him bringing in the air base....After it died down, they could visualize what was being done."

The Polish people of Chicopee, Springfield and Holyoke also banded together with Stonina's "Bonds for Bombers Committee" to make a four-county effort to sell enough war bonds, some $2.5 million, to raise funds for six B-24s. The *Liberators* were dedicated at Westover Field on Sept. 1, 1943, and were officially accepted by Brig. Gen. R.E. Nugent, chief of staff for First Air Force. The bombers were named *Polish Yankee Spirit of Franklin County, Spirit of Berkshire County, Spirit of Springfield, Spirit of Hampden County, Spirit of Chicopee* and *Spirit of Hampshire County.*

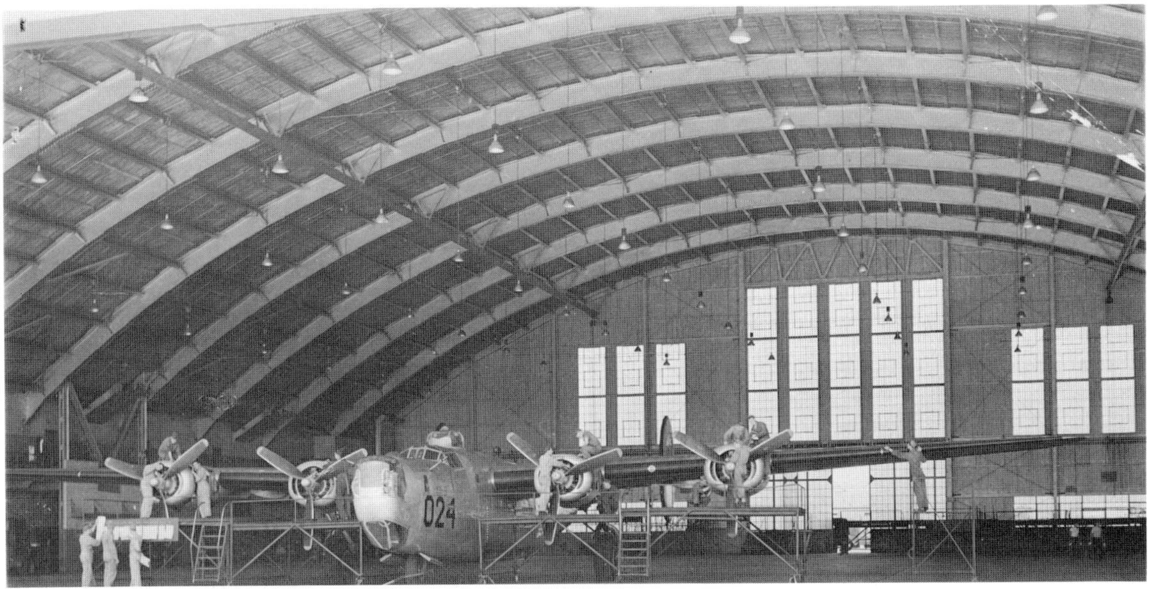

OFFICERS of the 806th Bomb Squadron gather in February 1944 alongside a B-24 Liberator parked in front of Westover's Hangar 9. The 806th was assigned to the 471st Bombardment Group which trained at the base from Jan. 28 until deploying to Europe on April 10, 1944. In top photo, a B-24 is overhauled in the Base Hangar at Westover.

AERIAL VIEW of Westover Field in February 1942 taken from the south looking north at 3,000 feet. The four tobacco barns in the lower left, were used by construction firms building the base. The main airstrip, Runway 19, was parallel to the four hangars.

The Army Air Forces had been gathering strength in Britain since the summer of 1942. On Aug. 17, 1942, Col. Frank Armstrong's 97th Bomb Group launched a dozen B-17s to strike railroad yards at Rouen, about 35 miles inside France. All returned 90 minutes later and none were damaged.

Seven other missions were flown in August of 1942 and no bombers were lost. By the end of the year, more than 1,500 bombing sorties were conducted with the loss of only 32 aircraft -- a loss of only 2 percent of the bombing force.

As combat experience was gained, the lessons learned were used to adjust skill training at various U.S. bases during the massive buildup of our air forces.

Westover Field had been training fighter pilots and bomber crews for anti-submarine warfare, but in late 1943 the emphasis changed and the base took young men from tech schools and molded them into combat air crews.

Once the young pilots, navigators, gunners and bombardiers -- mostly teenagers and men in their early twenties -- arrived at Westover they were formed into crews.

Their few months at Westover Field and their training missions over Western Massachusetts were vital to the American bombing plan.

Tremendously high pressure developed after the invasion of North Africa, and the Casablanca meeting between President Roosevelt and Prime Minister Churchill, for the American bombers to carry the war to Germany.

The Eighth Air Force -- now under Brig. Gen. Ira Eaker, a fighter pilot who had flown the mail for Oscar Westover -- began striking Germany in January 1943 and immediately suffered heavy losses.

When General Eaker arrived in Casablanca, General Arnold told him the U.S. would be joining the British in night bombing because Winston Churchill thought the Eighth Air Force could not sustain such heavy losses. Eaker later quoted Churchill as saying, "Now, you understand there's no lack of confidence in you. I greatly admire

352nd Fighter Group

The 352nd Fighter Group trained at Westover Field in P-47 *Thunderbolts* and later switched to P-51 *Mustangs* after joining the Eighth Air Force in England. The 352nd flew escort for 214 of the 254 bombing missions flown by the 401st Bombardment Group (H) which modestly pegged itself as the "best damned outfit in the USAAF."

Pilots of the 352nd were credited with destroying 38 Luftwaffe aircraft on Nov. 2, 1944, when the 401st bombed Merseburg. Only one bomber was lost that day. In addition to that day, which was the second highest record for a fighter group, the unit had 10 other double-digit days of knocking down enemy planes. On Aug. 6, 1944, the 352nd was credited with a dozen Luftwaffe aircraft, six of them shot down by Maj. George Preddy.

The former Westover fighter group lost 56 *Mustangs* during the war, but claimed credit for downing 310 German planes.

Mountaintop ceremony honors B-24 crash victims

"Thirty years ago I walked down the side of Mount Holyoke and into a story that has haunted my mind ever since." With those words, Wilbraham resident Frank W. Tencza begins an incredible story that began 45 years ago at Westover Field and ended in 1989 during a memorial service at Skinner State Park in South Hadley.

The 44-year-old Vietnam veteran spent three decades researching the crash of a Westover-based B-24 *Liberator* which crashed into Mount Holyoke during a night training mission on May 27, 1944. The Army Air Forces bomber crew, preparing for overseas deployment 10 days before the invasion of Normandy, had taken off from Westover's main runway three minutes earlier. Everyone aboard perished.

Tencza grew up in the shadow of the mountain and stumbled upon the crash site while hiking in 1958. What started out as boyish curiosity has evolved into a life-long crusade to honor the *Liberator's* 10-member crew. Years were required for him to learn the identity of the plane and the names of the aviators who died in the crash. Through sheer persistence and detective work that consumed thousands of hours, he also located and contacted surviving relatives of each of the crewmen.

Over the years, he visited the crash site frequently, often placing an American flag there as his tribute to the fallen fliers. He singlehandedly organized a ceremony in 1989, where he joined with the survivors to dedicate a memorial to the B-24 crew.

Many of the relatives had never learned the details of the tragedy; and several said the mountainside ceremony enabled them to complete a grieving process that had gone on since World War II. The 10 Army Air Force aviators were remembered on Memorial Day weekend 1989 on a summit overlooking Westover and the crash site. The ceremony culminated Tencza's 31-year quest.

That monument, carved out of Massachusetts granite and adorned with a nickel-plated B-24 propeller, was dedicated during the sometimes emotional ceremony at Mount Holyoke. Engraved on the face of the granite slab are the names of the B- 24 crew members: Sgt. Arnold H. Anderson, Lt. William M. Ashley, Sgt. Wilburn H. Deschert, Lt. Donald D. Dowden, Sgt. Ambrose D. Griffith, Cpl. Ronald C. Lloyd, Lt. John D. Logan, Lt. Talbot M. Malcolm Jr., Cpl. Robert J. Ohr and Cpl. Kearney D. Padgett.

Most of the crewmen were in their late teens or early 20s. All were stationed at Westover. "These men shouldn't be forgotten," said Tencza during the dedication, which featured flyovers by vintage PT-13 and PT-23 trainer aircraft, and by modern F-15 *Eagle* fighters from the Massachusetts Air National Guard.

Alluding to the beauty and peacefulness of the mountain, which overlooks the Connecticut River, Tencza pointed out the paradox..."That on one awful night so long ago, a night when the mountain was on fire and machine-gun cartridges were exploding, those young men became a part of forever."

Following the dedication of the memorial, which Tencza said is also a tribute to all World War II aviators who died flying from Westover, family members placed roses at the base of the monument while a bugler from the Westover Honor Guard played taps.

Sixty surviving family members of the crew attended the ceremony, some traveling from as far away as Colorado, Georgia, Mississippi and Virginia to attend.

THUNDERBOLTS alongside a tent of Westover's 19th Maintenance and Supply Squadron.

the operation of these gallant young men you are leading. But my mother was American, and it just breaks my heart to see these wonderful young Americans sacrificed." Eaker convinced Churchill that U.S. daylight bombing and British night bombing would give the Germans no rest. Churchill withdrew his objection to daylight bombing by the Eighth Air Force and the sacrifices continued.

During February 1943, for example, General Eaker received 63 replacement B-17s, but the Eighth had lost 75 on combat missions.

There were other operational problems, too. On an 86-bomber raid over Hamm, only 39 bombers were able to drop their bombload -- the others had bomb-bay doors that were frozen shut!

Planes damaged in combat or landing mishaps were also stacking up in England because there were not enough repair facilities or mechanics available.

At the uppermost levels of command there were second thoughts about the feasibility of the American bombing effort from England and a questioning of the priority of Europe over the Pacific theater.

Commanders of flying units in East Anglia said the Army Air Corps had to better prepare the fresh bombing crews for operations over Germany. More realistic training was needed.

Winter crew training at Westover Field took on new importance because of the European experience. The aircrews flying over Western Massachusetts inside their

359th Fighter Group

The 359th Fighter Group trained at Westover in P-47 *Thunderbolts* and later P-51 *Mustangs* from January to October 1943 prior to deployment to England. From RAF Field 133 in East Wrentham the unit's 368th, 369th and 370th Fighter Squadrons flew 346 missions from Dec. 13, 1943, to Apr. 20, 1945.

The former Westover unit was credited with officially destroying 373 Nazi aircraft and 23 probables, damaging 185 aircraft, destroying 364 locomotives, 1,000 boxcars, and 200 canal barges. The fighter pilots also strafed and destroyed hundreds of military vehicles, light tanks, troop carriers and German staff cars while also providing fighter cover for hundreds of former Westover crews flying in B-17s and B-24s.

unpressurized, open-waist B-17s had a taste of what flying out of East Anglia might be like. Thick, fleece-lined leather jackets, bulky gloves and oxygen masks caused door gunners great difficulty, yet for every three men wounded in combat, another four suffered frostbite.

The crew's navigator and bombardier were crammed into the nose compartment with life jackets, oxygen bottles and machine guns. The B-17 was a cramped and uncomfortable vehicle for 10 strangers to commute to war. A tightly knit crew was a necessity.

Westover provided the common ground for crews to work together and determine their own strengths and weaknesses. Crushed caps, nicknames, and colorful flight jacket art contributed to the esprit de corps developed at Westover.

Crews were "staged" or formed at the base, and if enough qualified crew members were not available, then individuals were shipped overseas to be replacements in existing crews. Most of the men hoped they would be able to find a crew at Westover rather than to be a replacement in a veteran combat crew which would later rotate without them.

"The mess halls and the Officers Club were open around the clock," said Ben Mathis, a former B-24 pilot. "Crews were coming and going at all hours of the day and night....Normally the average stay was five or six days. They would be sure all of the crews had all of the equipment they needed. They usually put them through the legal section to be sure all of their wills were in place."

Pilots were assigned to many hours in Link trainers and gunners were allowed to fire a few hundred rounds at the base, but most of the time was spent forming into crews and learning about the other eight or nine young men who would be commuting to war in the same fragile bomber.

> ***"I remember that terrible collision one Saturday afternoon over the Holyoke Stop & Shop plaza where one P-47 Thunderbolt landed in the residential area and the other one landed farther up toward the river."***
> ***-- Neil O'Leary***

Mathis said the crews were shipped out as soon as they were ready. "They were so desparate for crews in England and they were losing so many....they were losing them like flies."

Residents of the cities and towns around Westover did their part by inviting crews into their homes and providing rides to the Chicopee USO and the Butterfly Ballroom in Springfield. The underlying thought among residents and aircrews was that these young men might not survive over Germany.

As 1943 unfolded and the "numbers game" became more serious, *might not* evolved to *probably would not.*

Near the end of July 1943, six missions were flown in seven days with the loss of 100 planes -- and 1,000 airmen dead, missing, wounded or captured in a week.

Col. Curtis E. LeMay had left Westover by then and was in command of the 4th Bombardment Wing scattered at bases in East Anglia. On Aug. 17, 1943, he led 146 bombers to strike the Messerschmitt factory, which built 200 Me-109s a month, at Regensburg. When the B-17s arrived over Germany, only 131 were left and Colonel LeMay lost 15 bombers before reaching the target.

That same day, 230 bombers departed England for the ballbearing factories at Schweinfurt. Only 194 returned -- the

BODY ARMOR

Flak suits and helmets were in vogue at Westover Field in 1944. The armored vests were issued to crewmembers bound for England. But many of the young airmen complained the heavy suits were too cumbersome and hampered their movement in combat.

The flak suit was composed of overlapping plates of maganese steel cased in vests and aprons of canvas. They were designed to protect against wounds from spent bullets, low-velocity flak and 20mm shell fragments. Army Air Force statisticians reported in 1944 that 85 percent of all air combat casualties came from low- velocity shrapnel.

Approximately 40 percent of combat wounds received by bomber crewmembers over Europe were caused by flak, 40 percent by 20mm cannon shells, 10 percent by debris blown away from their own aircraft, and 10 percent by machinegun bullets.

Although the vests and aprons were only useful against low-velocity shrapnel and spent rounds, a study of 133 airmen hit over Europe while wearing the armor indicated that 65 percent received no injuries and 24 percent were only injured slightly. But 2.3 percent were seriously wounded and 8.2 percent were killed.

The M1 vest was armored front and back and weighted slightly more than 18 pounds. The M2 vest had armor only in the front and was worn by pilots in armor-plated seats. The M4 full apron was designed to be worn by gunners and weighed nearly eight pounds.

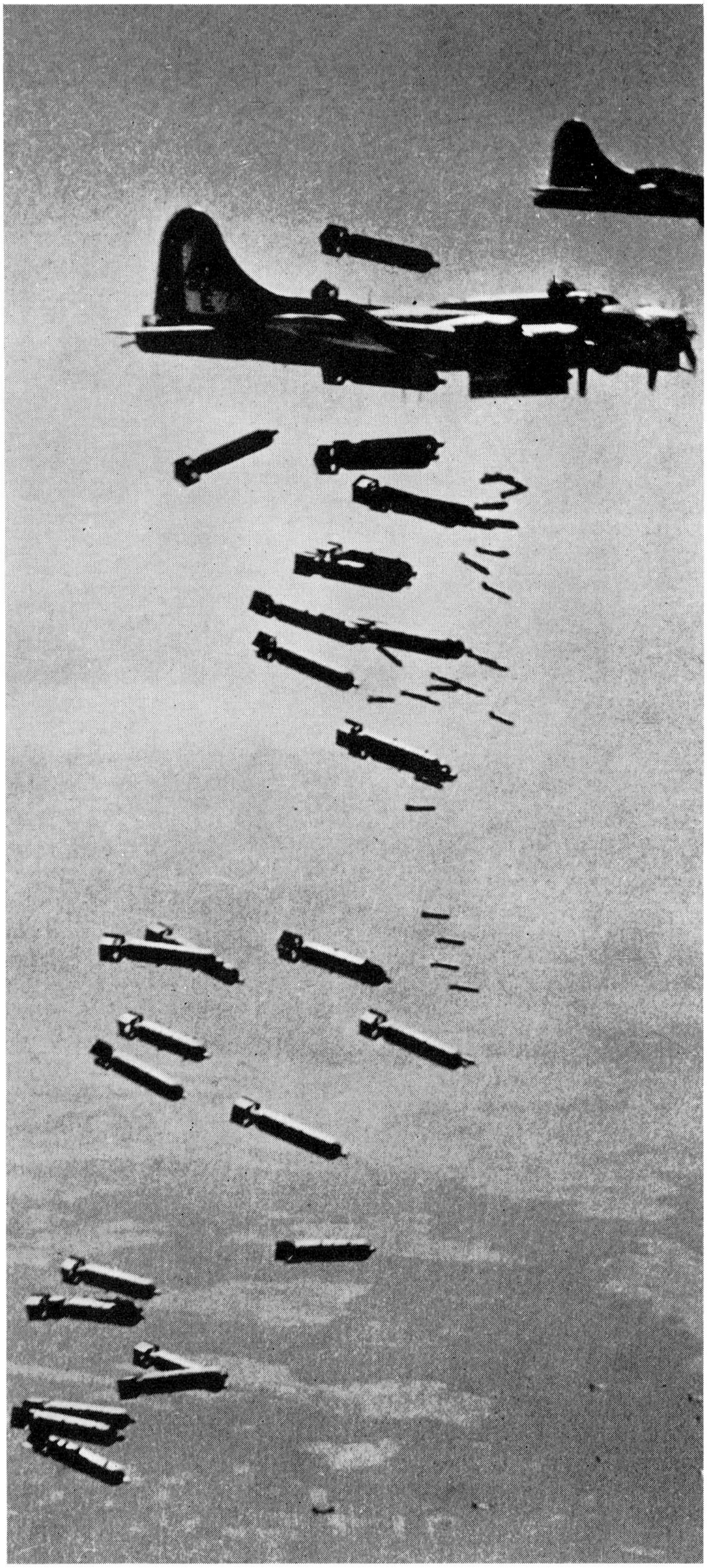

loss of 360 airmen -- and of the returning bombers, 27 were so badly damaged they would never fly again.

In three weeks, General Eaker's mighty Eighth Air Force had lost 147 aircraft. But when the aerial photographs were studied, the British were amazed at the accuracy of the American strikes, thanks to the Norden bombsight and aircrew dedication.

Schweinfurt was attacked again on Oct. 14, 1943, and of 291 bombers launched, only 227 reached the target and 60 bombers and 600 men were lost. Many of the returning bombers and airmen would never fly again.

Strategic bombing was feasible, as long as American industry and training bases like Westover Field could keep the planes and crews coming.

General Westover's decision to gamble on the B-17 and the efforts of men and women at the "super airbase" that bore his name were important factors in the European victory.

President Roosevelt and Prime Minister Winston Churchill met in Quebec during August 1943 and agreed on the outline for what was to become Operation Overlord, the Allied invasion of *Festung Europa*. British officers soon arrived at Westover Field to watch C-47s of the Troop Carrier Command tow Waco gliders and use airborne tactics which would be used in the invasion. Preparation for the invasion called for more coordination between the RAF Bomber Command and the Eighth Air Force and the creation of Allied Strategic Air Forces in the spring of

1944. The invasion support plans called for intensified bombing of the German heartland and heavy aircraft and crew losses were predicted.

By the end of the war, more than 2,000 air crews for B-17s, B-24s and B-26s were trained at Westover. During the first quarter of 1944, 125 crews were trained and departed for Europe. By July, another 724 crews had been trained at Westover.

Bomber crews from Westover flew to Newfoundland and then on to Heathrow in England for assignment to the many Eighth Air Force bases.

When flight training was frozen in April 1945, there were 289 trained B-24 combat crews assigned at the base. The B-24s were all replaced by B-26s on July 19, 1945.

"HELL'S ANGELS," an Eighth Air Force B-17 crew photographed with their ground maintenance and armament teams after surviving their 31st mission over Europe.

German prisoners at Westover

J.G. Roy and Sons construction company of Springfield was awarded a $25,000 contract in August 1944 to build a 250-man prisoner of war compound at Westover by mid-October. The POW camp, complete with sentry towers and guard shacks, was built in the mid-western part of the base between the main and industrial gates as far from the flight line as possible. The area was surrounded by chainlink fences, and 3- by 12-foot steel runway mats were buried 30 inches deep to discourage tunneling beneath the fences.

The first group of former Luftwaffe airmen arrived at Westover on Oct. 6, 1944. Since the POWs arrived too late to be helpful in the local autumn harvest, they were assigned to KP and other menial tasks at Westover. They worked in the salvage yard, on grounds duties, in the motor maintenance shops, and on plumbing details, plus some were contracted out for private labor projects. Within two months, Westover officials were asking the War Department to send more POWs to the base. The base historian wrote in late 1944 that "a great improvement in cleanliness, neatness and appearance can be seen. In fact, the use of these prisoners has been found to be of such value that another 250 are being requisitioned." The POW compound was designated a separate unit in March 1945 and more prisoners arrived at Westover the following month. Of the 523 POWs assigned, 190 were assigned to KP duties. There is no known record of any escape attempts or reported difficulties with the German prisoners of war.

"We used to slip them cigarettes," said Betty O'Connell, who worked in base supply in 1944 and 1945. "I kept thinking of my two brothers who were in the service, and I hoped that somebody would care for them if they were ever in the same situation."

When the war ended, the base rapidly coverted to a peacetime 40-hour week and the primary mission was the separation of personnel. The survivors among the thousands of young airmen were returned from Europe and processed at Westover Field for discharge.

FINAL FREE RIDE driver awaits grinning GIs who were discharged Oct. 18, 1945, from Westover Field's U.S. Army Air Forces Separation Center.

PRESIDENT HARRY S TRUMAN'S plane "the Columbine" undergoes major maintenance overhaul in Westover's Hangar 3.

AIR TRANSPORT COMMAND

Flights from Westover are an important link to Europe for 900,000 passengers

While still in the midst of processing thousands of Army Air Forces servicemen returning from Europe, Westover Field was tapped in January 1946 to become the principal port of entry and embarkation for the Air Transport Command's North Atlantic operations to Europe, Africa and the Mideast.

Westover was selected to gain the major ATC passenger and freight operations at LaGuardia Field, N.Y., and those at Wilmington, Del., and Grenier Field, N.H. Army Air Force operations in Iceland and Canada were also placed under the command of the Westover unit.

Brig. Gen. Haywood S. Hansell Jr., a Pacific Theater veteran, was named commanding general of the ATC's North Atlantic Wing. He predicted Westover Field would become one of ATC's greatest maintenance centers in the nation. A second major ATC base for South American filghts was slated for Morrison Field near Palm Beach, Fla.

Customs and immigration offices were established at Westover and the command scheduled an air shuttle service to Washington, D.C.

General Hansell said all Army Air Forces travel to and from Europe and Africa would be through Westover and he believed all military travel would someday be consolidated under the ATC.

The base was assigned 175 special five-man crews for the operations and 140 officers and 300 enlisted men assigned to the First Air Force at Westover were brought into the ATC.

Base Commander Col. Russell Keillor said 850 civilians employed at the base were assigned to the ATC and civilian recruiters were seeking another 1,500 skilled mechanics for the new C-54 operation.

Although glider tactics had been tested at Westover Field during the war, the arrival of the Air Transport Command was the true introduction of the base to strategic airlift. An introduction which would eventually develop into a major concept and be returned to Westover AFB.

The concept of massive strategic airlift took nearly four decades after the Wright brother's first flight to become institutionalized.

The top American thinkers on airpower, including Maj. Gen. Oscar Westover, were proponents of strategic bombing and the fighter support necessary for escort and the defense of forward areas.

Other nations moved rapidly in the post-World War era to develop tactical bombing and airlift, but the United States relegated the movement of low density, high value cargo to the emerging civilian airlines.

HOSPITAL PATIENTS from Paris land at Westover Field aboard an Air Transport Command C-54 after being evacuated from war-torn France.

The far-flung operations of World War II changed American thinking dramatically. Aircraft ferry operations were modeled on what had been learned by the airlines; and these operations were quickly modified again to establish the techniques for the world's first real airlift -- supplying China "over the Hump."

On May 29, 1941, the Air Corps Ferrying Command was established to move aircraft -- many in support of Lend-Lease to Britain -- and Westover Field was a major stop on the route to England.

By 1943, necessity in the form of Japanese closure of the Burma Road created aviation's first 24-hour-a-day strategic airlift to supply Chiang Kai-Shek's government in the battle against more than a million Japanese troops.

***KARACHI CAMEL CARTS** lug supplies to an Air Transport Command C-46 for delivery over "the Hump" to Fourteenth Air Force personnel in China.*

Off muddy 6,000-foot strips at Chabua in the upper Assam Valley of India, C-46 and C-47 pilots, along with aircrews flying in converted B-24 bombers known as C-87 cargo planes and C-109 tankers, flew with practically no navigation aids over the Himalayan range to Kunming, Chungking, and other fields in China.

Flying the 500-mile mission over "the Hump" were Flight Officer Service Pilots, civilian-trained -- and usually older -- airmen sporting a large "*S*" on their wings.

This initial cadre of former airline and crop-dusting pilots was not supposed to be used in combat, so the airlift concept was an ideal War Department mission.

More than 1,000 men were killed, and more than 400 aircraft were lost to Japanese fighters and rugged flying conditions. Some pilots and aircraft were lost due to inexperience resulting from wartime pressure to make airlift work.

"Flying 'the Hump' was like playing blackjack against a professional dealer; do it long enough, and you lose," said aviation writer Don Downie, a member of the Hump Pilots Association.

Mr. Downie says the men and their fantastic operation are still memorialized today on Indian maps. After an officer termed the men "dumb bastards," the center for Assam Valley operations is still known as Dumbastapur.

But the learning curve soared during the world's first real airlift. And the Army Air Force institutionalized airlift in the form of the Air Transport Command, which became jokingly known in the CBI Theater as "the Assam Trucking Company."

At the beginning, the ATC used a 50-mile wide corridor for two-way traffic between 18,000 and 25,000 feet. As operations expanded to an average of 650 trips a day -- a take-off every 200 seconds -- the airbridge was expanded to a corridor 200 miles wide.

"This was all new," wrote Lt. Gen. William H. Tunner, commander of "Hump" operations at the peak. "No other air operation -- civilian or military -- had ever attempted to keep its fleet in continuous operation all around the clock, in all seasons, and in all weather."

UPPER ASSAM field was a quagmire during the monsoon season of 1943 when this ATC crew walked from their C-46 Commando.

When the massive ATC operation ended in 1945, some 237,572 cargo missions had been flown over "the Hump" to deliver more than 650,000 tons of war material.

Some 1.2 million passengers were carried, along with nearly 200,000 combat loaded Chinese and American troops and nearly 5,000 pack animals.

Another 4,720 military patients were air-evacuated to hospitals and ships in India.

Winston Churchill termed the airlift feat as "...astounding efforts at a vast cost." The prime minister told the House of Commons, "Certainly no more prodigious example of strength, science, and organization in the class of work has ever been seen or dreamt of."

At the mission's peak, there were 722 aircraft involved and more than 85,000 men assigned to the operation.

The Fourteenth Air Force, today Westover's parent Air Force Reserve unit at Dobbins AFB, Ga., was one of the prime users of the equipment, supplies and ammunition flown over the Himalayas.

In January 1945, the Fourteenth "Flying Tigers" recorded 334 Japanese aircraft destroyed, 48 probably destroyed, 215 damaged, and 13,500 tons of shipping sunk with another 58,900 tons damaged and probably sunk.

To support the Fourteenth Air Force, the ATC's C-46s carried a cargo load of 8,000 pounds with full fuel or 11,500 pounds on short tanks. On a single day in January 1945, there were 500 fully loaded missions to China -- a take-off every 75 seconds.

The roots of the current strategic airlift mission at Westover are deep in the mud of the Assam Valley. Compare the brave efforts of these early pioneers flying C-46s crammed with 8,000 pounds with the modern capability of Westover's 439th Military Airlift Wing with a single C-5A lugging 250,000 pounds.

A year later, in January 1946, the ATC announced Westover would be the new headquarters for the command's North Atlantic Wing.

And two years later, General Hansell's predicition of consolidated airlift would come true -- the Military Air Transport Service was formed and operated by the new U.S. Air Force.

NATIVE LABOR and American rock-crushing machines enabled ATC to develop the Assam Valley airfields in 1944 to withstand the heavy usage of airlift operations. As the U.S. learned the value of airlift and committed more resources to air cargo operations, the missions to China became safer and more productive. Many of the lessons learned during "the Hump" operations were later put into play during the Berlin Airlift and have been modified for use at Westover AFB today by the 439th Military Airlift Wing.

OPERATION VITTLES 1948

Westover becomes the U.S. hub to support the Berlin Airlift

In 1948, Westover Air Force Base was the U.S. hub for overseas flights supporting the costly airbridge keeping the free sector of Berlin from Soviet occupation.

Greater Springfield residents wanting to demonstrate support for the tremendous Cold War effort to retain Berlin and avert war soon found an opportunity through a voluntary campaign in Western Massachusetts.

Westover became a Military Air Transport Service base on Feb. 1, 1948 and aircraft stationed at the base ceased all domestic flights. Regular flights out of Westover were scheduled for Greenland, Goose Bay, Labrador; Harmon Field, Newfoundland; and bases in Europe and Africa. All northern bases were dependent on Westover for transportation of the majority of their supplies, replacement personnel and fresh produce.

Four days later, the Air Transport Service and the Naval Air Transport Service officially set up the combined air system known as MATS which was operated by the Air Force. General William H. Tunner, the father of modern airlift because of his experience running "the Hump" operation over the Himalayas, became deputy commander of MATS and assigned Capt. Robert Hogg as his Atlantic Division administrative officer at Westover.

Since V-E Day, relations among the Allied superpowers and the Soviet Union had been in decline. Berlin, deep in Soviet-controlled Germany, was occupied by the four powers and supplied from the West by highway, rail and canal.

On June 18, 1948, West German currency was reformed to aid the economic miracle of the Marshall Plan. The following day, Marshal Sokolovsky, the Soviet military governor of Berlin, issued an angry statement terming the city "part of the Soviet occupation zone."

Russian soldiers began tearing up railroad tracks, marking bridges as unsafe for vehicular traffic and putting obstacles to navigation in the canals.

The Allied leadership estimated the city could hold out without resupply for only 30 days.

When the Soviet blockade was complete on June 24, 1948, Berlin's mayor-elect, Ernst Reuter, orchestrated a mass gathering of 80,000 people. His call for united defiance of the Soviet blockade was answered by a roar of approval.

Short of possibly starting a war by having an armored column drive to Berlin, the Pentagon's only available non-combat option of a massive skybridge to Berlin was preferable to fighting an other war.

The airlift lessons learned a few years before over the Himalayas were honed and combined with ideas gathered from

MILITARY AIR TRANSPORT SERVICE C-54s are unloaded at Templehof Airport in Germany. The Berlin Airlift was the first action taken by MATS and the new U.S. Air Force during the Cold War.

WILLIAM H. TUNNER
The father of modern airlift

strategic bombing and matured into a professional concept of airlift.

Soon after General Tunner was tapped to run the operation he flew to Westover AFB and picked up Captain Hogg, who became one of his major task force planners. As the general assembled a planning team of former ATC "hump" pilots, they realized quickly that resupplying the military force in Berlin would be relatively easy.

The severe problem was providing food, fuel and medicine for nearly 2.5 million civilians in West Berlin, a city which before the blockade had imported 15,500 tons a day.

MAJ. GEN. CURTIS LEMAY, 1948

"Can you transport coal by air?" the city's American military commander asked the USAFE chief, Maj. Gen. Curtis E. LeMay, the former operations officer at Westover.

"Sir," General LeMay reportedly responded, "the Air Force can deliver anything!"

What was to become world famous as "Operation Vittles," began on June 27, 1948, with just 102 C-47s, each capable of lugging 6,000 pounds, and only two C-54s with 10-ton capacity in all of Europe.

In the first two days, 80 tons of flour, milk and medicine were delivered. The 1,000-ton mark was reached by July 7, 1948. People began wondering if airlift could even sustain a city geared to importing more than 100,000 tons a week.

The months-old Military Air Transport Service had less than 300 C-54s in the entire system, and many of these planes were heading to Westover AFB to become part of the airbridge to Berlin.

Newspapers and newsreels carried reports of haggard pilots sleeping under desks in Base Ops, and staff officers flying

EXHAUSTED AIRCREWS unload supplies from one of the many C-54s arriving in West Berlin to aid the nearly 2.5 million civilians there.

aircraft, but to General Tunner these heroic news accounts were just indications of an amateur airlift doomed to failure.

"The capability of airlift was unknown to Europe," he wrote in 1964, as the concept "was generally unknown in our military...In air transport everything is different -- rules, methods, attitudes, procedures, results."

In India, there had been Dumbastapur and a dozen other bases feeding into a half-dozen Chinese fields. In Germany, the Allies had only three air corridors to Berlin's two airfields -- Gatow in the British sector and Templehof in the American portion.

Built in 1935, Templehof was an architectual wonder with a whole Messerschmitt factory underground, but the sod runways would never withstand the around-the-clock punishment from fully loaded C-54s and C-47s. One runway had steel mats -- forerunners of PSP -- over rubble, and Allied sappers scurried to make repairs between air operations.

During the first month, ground operations were being thwarted by exhausted and hungry aircrews walking away from their aircraft during unloading and refueling. The German Red Cross solved the problem by sending out mobile snack bars staffed by their most beautiful volunteers. "There were no more

GERMAN CHILDREN at Templehof create a sandbox version of the Berlin Airlift.

moans about staying by the plane," General Tunner said. "You couldn't chase them."

By August of 1948, turn-around time was down to 30 minutes and an aircraft was landing every three minutes. But heavily laden planes were crashing, others were stacked up in the limited airspace, air traffic controllers were going nuts with frequen cies jammed with chatter, Soviet fighters were buzzing and playing "chicken" with aircraft in the narrow corridors -- there was still much to be learned about airlift.

To maintain the flow, General Tunner laid down harsh groundrules. Any pilot who messed up an approach had to exit the airspace and fly 200 or 400 miles back to the base of origin for rescheduling.

Any pilot who failed to land with more than 400 feet ceiling and a mile visibility would be reduced to copilot. Any pilot landing with less than 400 feet and a mile would be court martialed. The RAF reduced limits at Gatow to 200 feet and a half-mile visibility.

Tales of tired pilots might get good press, but lousy flying wasn't going to avert war and keep Berlin in the Allied camp.

A young Air Force lieutenant flying missions into crowded Templehof Airport in the U.S. sector of Berlin felt sympathy for the children waving to the American aircraft.

UNCLE WIGGLY WINGS, Lt. Gail S. Halvorsen, launched "Operation Little Vittles," when he began airdropping candy to the children of Berlin.

Lt. Gail S. Halvorsen of Utah was one of the many Westover pilots flying "Operation Vittles" supply missions into the divided city of 2.5 million. As a C-54 pilot, he was flying three 90-minute missions over Soviet-controlled territority into Templehof, breaking for six hours of crew rest, and then flying three more missions.

With General Tunner cracking the whip and professionalizing airlift, young airmen like Lt. Halvorsen had little contact with Berliners other than those on the tarmac.

One day when Lt. Halvorsen was not scheduled to fly airlift missions, he caught a hop into the city. "I was taking pictures of other guys coming into the airport and there were some kids watching from outside the barbed wire fence around the airfield," he said.

Some of the German waifs spoke English, but not one of the highly disciplined children asked the American pilot for candy.

Lt. Halvorsen had only two sticks of gum in his pockets. He remembers breaking them in two and handing out four pieces.

"Their reaction was remarkable," he said, "and it was then that I realized I wanted to get them more....so I told them I'd drop some on my next flight."

He told the children he would wiggle the wings of his Douglas C- 54 *Skymaster* so they'd know which aircraft would be dropping candy.

The legend of "Uncle Wiggly Wings" or the "Candy Bomber" began that night as Lt. Halvorsen created tiny parachutes from handkerchiefs and attached small bags of candy bars, gum and lollipops.

The next day, "Operation Little Vittles" began as the young pilot from Utah began rocking the wings of the C-54 as crew members dropped candy down the transport's flare chutes.

Word of the unauthorized candy parachute drops soon spread among the American aircrews and the cluster of children outside Templehof increased daily.

The word also climbed the chain of command. The young Westover pilot was told to report to his wing commander for what he expected would be a reprimand.

LANDING AT TEMPLEHOF

General Tunner was, of course, trying desperately to build on the "Hump" experience to professionalize the Berlin Airlift, yet as daily tonnage was increasing more pilots were coming in low and slow on final approach with 10-ton loads and making candy drops. From a command vantage point, "Operation Little Vittles" was unprofessional and possibly dangerous, yet this purely American response was the essence of the spirit of the Berlin Airlift.

Instead of a reprimand, Lt. Halvorsen received a pat on the back.

The news media picked up the "Little Vittles" story and articles appeared in the U.S. He was sent back to the states briefly for an appearance on a popular nationwide radio program, "We the People" and he visited Chicopee.

Soon candy and handkerchiefs began pouring in to Westover and other bases for shipment to the West German airfields supporting the Berlin Airlift.

A candy manufacturer began shipping cartons and volunteers in Greater Springfield were crating candy for shipment to Lt. Halvorsen's 17th Air Transport Squadron at Rhein-Main.

CREW KILLED in August 1948 after brakes of an ATC C-54 failed and aircraft skidded off the runway at Templehof.

Chicopee Mayor Edward Bourbeau spearheaded the idea of making his city the U.S. headquarters for the candychute program. City officials soon set up a committee of 50 children representing the city's 22 schools and set up operations in an abandoned fire station at the corner of Grape and Springfield streets.

The Operation Little Vittles Committee was incorporated and began official operations on Oct. 14, 1948, at Chicopee High School. Donations began pouring in from around the country.

Initially, 2,000 custom stenciled boxes were donated by New England Container Co., 2,000 sheets of cloth came from Chicopee Manufacturing Co., and the Veterans of Foreign Wars from Connecticut donated 3,000 handkerchiefs for parachutes. Thousands of rolls of *LifeSavers* were donated by the New York candy manufacturer and 11,000 yards of ribbon came from Budd Co. of Pennsylvania.

Students at Elms College and in Springfield and Chicopee schools set aside time to prepare the candychutes. Cloth had to be cut, corners tied with string or ribbon and candy bars or rolls of *LifeSavers* attached before delivery to Westover AFB.

PROTEST RALLY on Sept. 9, 1948, against the blockade of Berlin. More than 300,00 Berliners gathered at the Reichstag.

When the Little Vittles operation finally ended, an estimated 100,000 little parachutes were tied to candy at the former Engine Co. No. 4 fire station in Chicopee.

Some 36,000 pounds of candy were flown out of Westover AFB for parachute drops over Berlin.

"We quit dropping on the end of the runway because it got too crowded, so we picked out the best places in the city to make drops and marked those on a big map," said Lt. Halvorsen.

"Before heading out to the aircraft, pilots would pick up their box of candy deliveries and mark on the board where they would make the drop," said the retired colonel, who ended his 30-year Air Force career by commanding Templehof Air Base.

While the now-authorized candy drops added spirit to the airlift operations and provided a means for American civilians to show their support, General Tunner continued to hammer away at professionalizing Operation Vittles.

One aspect needing attention was international cooperation. The French could commit little to the Berlin Airlift

AIRBRIDGE SKYTRAINS unload at Templehof.

because most of their transport aircraft were involved in Indochina. The British had an odd mixture of World War II aircraft, but their flying and support operations were much different than the American concept.

Yet the two RAF bases in West Germany were an hour closer to Berlin than Rhein-Main AB, so two U.S. aircraft at Celle or Fassberg could do the work of three at Rhein-Main. General Tunner assigned American crews to the RAF bases.

Loadmasters and pilots were concerned with calculating the dif ferent weight mixtures with imperial gallons and British long- tons of 2,260 pounds, plus metric liters and kilograms, that constituted the daily loads which were growing beyond 9,000 tons a day to supply a city geared to using 15,000 tons a day.

One day, General Tunner saw civilians loading coal sacks marked for 100 pounds, but the hardworking men only weighed 1 percent of the sacks.

When he checked, the Allied airlift commander discovered some jute bags ran up to 125 pounds and the average was 115 pounds.

"A fifteen percent overloading that could mean a ton and a half extra weight on a C-54 full of coal," the general said. "No wonder the plane seemed sluggish." He called in the Army to weigh trucks lugging goods to the aircraft.

THREE CREWMEN DIED on June 25, 1948, when this C-54 crashed into an apartment complex in West Berlin.

Coal and flour dust created a maintenance problem, until an ingenious American mechanic dropped a rubber hose out a porthole and used the slip stream suction to vacuum the plane during the return flight.

Berlin needed 38 tons of salt a day, but the corrosive nature of spilled salt was eating away at aircraft alloys and cables. The British had Sunderland flying boats treated to withstand salt water, but the big planes could only land on Berlin's Havel Lake. Before the lake froze in late 1948, the RAF designed special containers and flew salt in the bomb-bays of Halifax bombers.

Toting 646 tons of food a day into Berlin caused American food processors to pay more attention to dehydrated products and more commodities were squeezed or boiled to remove water. Packaging and dehydration had made great strides during World War II, but much of the bulk supplies needed in Berlin were different than the commodities prepared for military use. Many of the lessons learned 40 years ago became the basis for procedures used today by Westover's aerial port squadrons and loadmasters.

As the Berlin Airlift matured, the 800 Allied aircraft were scheduled by simple arithmetic. With 1,440 minutes in a day and aircraft spaced at 90 seconds between each take-off or landing, there could be 480 landings at each of the airfields.

In the three 10,000-foot, 20-mile wide corridors, aircraft were stacked at four levels in each direction, at 500-foot intervals. At each level, the aircraft were 15 minutes apart, but since there were four streams in each direction, the airflow in the corridors had the transports separated by 1,000 feet and three minutes.

ONE MILLIONTH TON arrived at Gatow airfield, Berlin.

There were mishaps from heavily loaded aircraft skimming five-story apartment blocks to approach Templehof. In midsummer 1948, a fully loaded C-47 strayed off course in fog and crashed into an apartment building. When fog increased in late 1948, the Soviets began plaguing Allied aircrews by releasing barrage balloons.

By the winter of 1948, coal was rationed at 25 pounds per household for the entire heating season. Although tons of coal were being flown into Berlin each day, nearly all was reserved for production of gas and electricity.

The greatest push came on Easter Sunday -- April 17, 1949 -- when 12,941 tons of coal -- enough to fill 600 freight cars -- were flown in 1,398 flights without mishap.

"It was that day, that Easter Sunday, I'm sure, that broke the back of the Berlin blockade," said General Tunner. "From then on we never fell below 9,000 tons a day; the land blockade was pointless."

On May 4, 1949, the Soviets agreed to lift the blockade and the Allies announced, "All communications, transportation and trade restrictions will be removed on May 12."

Fearing the Soviets might be opening ground traffic just to get the Allies to dismantle the massive airbridge, the Berlin Airlift was continued for another three months to assure a stockpile of vital supplies.

On Aug. 12, 1949, the first of the Westover C-54s on loan to the Berlin Airlift was released and arrived back at the base for regular MATS missions.

On Sept. 1, 1949, Operation Vittles ended.

The creation of NATO and the eventual construction of the Berlin Wall both have roots in the massive Allied airlift.

The official cost estimate was $300 million and the loss of 72 people, of whom 35 were American. There had been 276,926 flights to haul 2,323,067 tons into Berlin.

Next to averting war, the most important aspect of Operation Vittles was the impact airlift made on strategic planning. Big planes with ramps for fast off-loading seemed to be the answer for future operations to maintain the peace or support combat operations. America may have been at peace, but the prospects for war showed the need for an ever ready airlift capability.

Some far-reaching strategic planners were even looking at the French experience in Indochina, the British in India and the U.S. in the Philippines and predicting that future Allied operations might not

OUTBOUND CARGO was limited from Berlin, but ATC crews carried a daily average of 2,500 tons of goods produced within the blockaded city.

have as many airfields available for forward operations. Big, long-distance aircraft seemed to be the answer for the emerging post-colonial era.

After the Berlin Airlift closed, General Tunner said, "The future of air cargo is not necessarily in high speed, it's in big air planes driven by powerful but economical and reliable engines. The big, efficient, dependable piston engine right now looks like the modern cargo plane power-plant answer, at least for some time."

The Berlin Airlift demonstrated the need for huge cargo carriers to support U.S. operations. The load of 899 C-47s could be carried by 178 C-54s or by only 68 of the brand new C-74 *Globemasters.*

General Tunner envisioned a day when a few huge inter-continental aircraft -- designed for cargo and not modified troop carriers -- could carry the needed goods.

THE FIRST CONVOY through the highway corridor into Berlin arrived just after midnight on May 12, 1949. The first vehicle was a British jeep that entered the checkpoint at Helmstedt.

At Westover AFB today, a C-5A *Galaxy* of the 439th Military Airlift Wing can carry more than 200,000 pounds, some 100 tons. To meet Berlin's minimum 4,500 ton daily requirement, America's C-5s could have flown only 45 missions a day.

Galaxies could have delivered the entire mission load of Operation Vittles in only 23,230 flights instead of more than a quarter million flights over 432 days.

FAIRCHILD C-119 in front of Westover's Base Hangar in the late 1940's. The Air Force Reserve was created in 1948 and during the Korean War many Reserve aircrews flew C-119s.

The AIR FORCE RESERVE

America creates a formidable flying force as the active duty armed forces are reduced

The creation of a separate Air Force in 1947, with a part-time component known as the Air Force Reserve, allowed much free-wheeling thinking unencumbered by the long-standing traditions of the Army and Navy.

The vision of General Oscar Westover had set in place a viable air arm on the eve of World War II. His former assistant, General Henry "Hap" Arnold, had led the massive expansion of the Army Air Forces into a major force for victory. Air Force leaders argued that technology and the reaction time necessary for successful air operations required a different mode of thinking. As General Arnold often said, World War I had been a test of brawn, World War II was a war of logistics, and World War III would be a war of brains.

Although a boost was gained from German technology harvested after the war, the slim Air Force budget had slowed drastically the quantum leap gained during World War II.

The brand new Air Force was only 400,000 strong in the late 1940s, only a sixth of the wartime peak. In the event of a national emergency, additional manpower was to be drawn from the Air Force Reserve and the Air National Guard. And, hopefully, the Pentagon planners would be able to avoid the problems with the reserve forces following World War I.

Back in May 1917 the First Reserve Aero Squadron was formed and the following month was assigned under the National Defense Act to the Aviation Section of the U.S. Army Signal Corps. At that time the aviation reserve corps was authorized 2,300 officers and men.

The next month kicked off America's serious effort in the air. On July 24, 1917, President Woodrow Wilson signed the Aviation Act of 1917, which budgeted a phenomenal $600 million for military aviation. The huge increase bolstered America's aviation industry and brought Oscar Westover to the rank of temporary colonel to oversee de Havilland DH-4 production.

AERO RESERVE de Havilland DH-4

Shortly after the United States entered World War I, two Reserve Aero Squadrons were ordered to active service with the Signal Corps. On May 24, 1918, some six weeks after U.S. pilots downed their first German aircraft, the Army removed aviation from the Signal Corps and formed the U.S. Army Air Service.

During the war, approximately 10,000 pilots were graduated from civilian and government training schools and were rated as reserve military aviators in the Air Service.

In the wake of World War I there was massive demobilization of the 195,023 men in the Air Service, and the Air Reserve nearly disappeared.

Two decades later, on the eve of World War II, there were only 23,455 men and a few hundred planes in the U.S. Army Air Corps. The service was ranked sixth among air forces in the world. In the small Air Corps were many reservists -- 1,500 Army Air Corps pilots, 1,300 non-rated officers and 400 enlisted men -- on extended active duty.

During World War II, the Army Air Forces grew to 2,372,292 men and women operating nearly 80,000 aircraft around the world. But when the war ended, there was tremendous public and political pressure to demobilize along the lines of the earlier post-war era. Westover Field was a major out-processing center for thousands of aviators and airmen returning from Europe.

By the end of 1945, the Army Air Force was down to 890,000 men and women. By the end of 1946, there were only 450,000 in uniform and even though Westover Field was a major installation as headquarters for the ATC's North Atlantic Wing, there were only a few thousand men and women stationed at the base. By 1948, the new Air Force's ranks were thinned to 305,827.

Faced with thousands of qualified aviators being mustered out at the end of the war, the Air Defense Command offered a postwar Air Reserve program which was merely a flying club. Qualified military aviators could train in a number of surplus aircraft. And surplus U.S. aircraft around the world were being sold cheaper than automobiles. Sturdy C-47s went for $500 and sporty P-51s for just a few thousand dollars.

Nine months after the U.S. Air Force was formed, on April 14, 1948, the U.S. Air Force Reserve was officially organized. An executive order from President Harry S Truman in October solidified the Reserve program. The Air Force placed the new AFRES program under Lt. Gen. Elwood R. Quesada as chief of the Office of Special Assistant to the Chief of Staff for Reserve Forces.

General Quesada had an outstanding record from commanding fighters in North Africa and Europe. He was also adept at public relations and was credited with improving relations between General Eisenhower and the Air Forces. After the war ended, an attempt was made to make Quesada the military aide to President Harry S Truman so the Air Forces could have direct access to the Oval Office.

In 1949, the Continental Air Command managed the national AFRES program of five light bombardment wings and 20 troop carrier wings, all training in units for immediate activation if a national emergency arose. If there was to be another Berlin Airlift, major Military Air Transport Service installations such as Westover AFB would be used for the units called to active duty.

GENERAL ELWOOD R. QUESADA in the cockpit of a P-38 Lightning. As a lieutenant, Quesada was a crew member of the Question Mark. During World War II, he commanded fighters. At the end of hostilities, Generals Arnold and Spaatz tried to have General Quesada assigned as Truman's military aide, but the Army preferred a ground officer. He became the top intelligence officer before the Air Force Reserve assignment.

When North Korean troops crossed the 38th Parallel on June 25, 1950, the 411,277-man U.S. Air Force was once again in disarray, making the transition from reciprocating engines to jets. Although the invasion of Korea came less than nine months after the Berlin Airlift, America's airlift capability was far less than adequate.

The Continental Air Command was called upon to provide Air Force Reserve members to supplement America's airlift capabilities.

The 439th Troop Carrier Wing (Medium) -- today, Westover's 439th Military Airlift Wing -- and two dozen other Air Force Reserve wings, about 30,000 members in all, were called to active duty for Korea and about 118,000 individual Air Force Reservists were recalled for active service.

But 15 of the 25 AFRES wings were quickly disbanded and members were reassigned throughout the again-growing Air Force instead of serving with the friends and teammates they had trained alongside. About 46,000 members of the Air National Guard were mobilized and flew aircraft left over from World War II.

But the tactics and war machines of World War II were soon overshadowed by technology. The new Strategic Air Command's gleaming silver state-of-the-art B-29 *Superforts* of 1945 were quickly outgunned by MiG-15s, which were 100 mph faster than F-80 *Shooting Stars*. The Air Force adopted British bomber tactics from World War II and black paint was applied to the B-29 undersides so they could be used at night. B-26

SAC's GLEAMING SILVER B-29s were quickly painted with black undersides to survive in the combat environment of Korea.

SABRE CREWMAN removes a gun panel from an F-86 at Osan Air Base during the Korean War.

Invaders were adapted with side-looking radar and found new life as convoy killers. The redesignated F-51 *Mustang* was the fighter workhorse in the early stages of the war, but jet fighters were on the horizon.

In February 1949, only 17 months before the Korean War began, the Air Force had received the first of 4,853 jet fighters designated the F-86 *Sabre*. The first 19 F-86A *Sabres* arrived

CHINESE MiG-15 as seen through the gun camera of Capt. Manuel "Pete" Fernandez' F-86 Sabre. Fernandez knocked down 14 MiGs and was the 26th ranking USAF ace of the Korean War.

in Korea nearly six months after the war started, and began combat air patrols on Dec. 16, 1950.

The following day, a North American F-86A scored the first of many kills against the Soviet swept-wing Mikoyan-Gurevich MiG-15.

With air superiority nearly assured, the Air Force concentrated on tactical bombing since there were few strategic targets left in Korea after the opening days of the war.

Tactical airlift was accomplished mostly by the Fairchild C-119, a radical twin-boom replacement for the venerable C-47 and the larger C-46. Evolved from the C-82, the new "flying boxcar" -- or "flyin' coffin" -- sported clamshell rear doors for the rapid aerial delivery of palletized cargo by parachute.

The C-119 could carry 15,000 pounds of cargo or 62 troops. The Air Force acquired the Fairchild C-119 in 1949, too late for the Berlin Airlift, but eventually 1,052 were purchased.

FLYING BOXCARS of the 61st Troop Carrier Group, "the Green Hornets," over Korea in October 1952.

Twin-engine Curtis C-46 *Commandos* were used to ferry cargo and troops between Japan and Korea.

Strategic airlift across the Pacific was a long process and the Air Force again turned to General Tunner to run the transport operation.

From the West Coast, Korea was 30 hours away by C-54 along the Great Circle route through Alaska and 34 hours away by the mid-Pacific route. By the end of the war, MATS delivered 80,000 tons of cargo and 214,000 troops across the Pacific while other transports carried supplies and people from Japan to Korea. Coming right on the heels of the Berlin Airlift, the Korean operation secured strategic airlift as one of the primary missions of the U.S. Air Force. In 1958, General Tunner became commander of MATS.

Aircraft were also need for medical evacuation from Korea and hospitals in Japan. Westover's hospital treated more than 2,500 Korean War casualties during November and December 1950.

Many lives were saved during the Korean War by aeromedical evacuation and the lessons learned in the early

1950s resulted in thousands of more lives saved a decade later in Vietnam.

The largest mass drop of paratroopers took place on Mar. 22, 1951, when 135 C-119s and C-46s dropped the famous 187th Regimental Combat Team at Munsan.

Korea also became a testing ground for experiments in aerial warfare as the helicopter was introduced for medical evacuation and old Grumman F6F *Hellcats* were refitted as pilotless flying bombs used in a manner not unlike Cruise missles of today.

One of the basic Air Force problems in Korea was tactical and strategic airlift. The South Korean runways were jammed, and the captured runways in North Korea were damaged.

PARATROOPERS of the famous 187th Regimental Combat Team board Fairchild C-119s on Mar. 22, 1951, for the largest airborne operation of the Korean War. Some 135 C-119s and Curtiss C-46 Commandos were used for the mass drop on Munsan.

Although Army engineers and Navy Seabees had learned much during World War II about airbase construction, the Pentagon realized the instant war in Korea would probably become the norm for future conflicts as the Allied powers sought to contain "the communist menace."

Although the U.S. had substantial bases in nearby Japan, the United Nations force had great difficulty transporting the supplies necessary to sustain the ground and air forces.

One breakthrough was the development by Douglas Aircraft Co. of the C-124 *Globemaster II*, a huge two-story four-engined transport. After extensive trials of a single "Air Proving Ground" machine in 1951, two squadrons were formed and the C-124s were deployed to Japan and Korea.

MATS eventually received 243 of the 300-mph C-124s which could carry 200 troops, 127 litters or 68,500 pounds of cargo.

The last C-124 *Globemaster II* was retired in 1975.

By the end of the Korean War, the Military Air Transport Service had carried 670,000 tons of material to Korea. MATS aircraft had also hauled 2.7 million passengers and evacuated 325,000 wounded to hospitals in the U.S. Many men from New England were sent by specially fitted Boeing C-97C *Stratofreighters* to the hospital at Westover AFB. The first Korean War casualty to be treated at the Westover hospital arrived in July

WESTOVER'S 1600th Maintenance Squadron repaired and serviced many MATS aircraft bound for Korea. Squadron members and civilian mechanics posed in 1952 near the Base Hangar before a transient C-124 Globemaster II they had serviced.

1950. As the war intensified that first year, Westover was receiving more than 1,200 wounded men a month.

But the big breakthrough in airlift was a Lockheed miracle -- the C-130 *Hercules*. The "Herk," which was later to play such an important role at Westover and in the Air Reserve Forces, was an amazing production aircraft developed in rapid time.

Learning lessons from Korea, the Air Force Reserve was reorganized in January 1955 into 24 wings and the Air Guard into 26 wings. The AFRES units operated 13 troop carrier wings, nine fighter-bomber wings and two tactical bombardment wings. Three years later, the Air Reserve Technicians program was adopted to keep AFRES units combat ready.

One of the new units was the 905th Tactical Airlift Group formed in 1958 at Bradley Field to operate C-119s. By 1965 the group was scheduled to receive C-124 *Globemaster II*s and both the unit and the airfield had outgrown each other. The 905th was assigned to Westover AFB as the base's first Air Force Reserve unit.

The 800-man group later converted to C-130s and was eventually absorbed into the 439th Tactical Airlift Wing and the 337th Tactical Airlift Squadron at Westover.

THE C-130 HERCULES was the miracle airlifter developed because of the Korean War. This C-130 is producing Spencerian rings of condensation as the turboprops bite into humid air prior to take-off from Pope AFB in North Carolina.

The Air Force Reserve was disrupted by the Korean War experience and years were required to rebuild the concept. Congressional interest in the aftermath of the Korean mobilization paid positive benefits for the reserve concept. Three major acts were legislated by Congress: the Armed Forces Reserve Act of 1952 standardized pay and training categories for the ready, standby and retired reserve; the Reserve Officers Training Act of 1954 created a promotion system; and the Reserve Forces Act of 1955 doubled the size of the Ready Reserve, allowed direct recruitment of non-prior service candidates and imposed sanctions for unmet training obligations.

The chief of staff of the Air Force announced in January 1955 an air reserve forces mobilization requirement for 51 wings -- 26 in the Air National Guard and 24 in the Air Force Reserve. By the end of the year, AFRES had 13 troop carrier wings, nine fighter-bomber wings and two tactical bombardment wings. Two years later, the first three of five air rescue squadrons were activated.

The mid-Fifties Reserve program also included training for individual reservists and non-flying units such as aerial ports, communications, maintenance, engineering, supply and medical units along with detachments for postal, censorship, weather and other tasks.

C-123 PROVIDERS became one of the workhorses of the Air Force Reserve after Korea.

But the major alteration of the Reserve concept was in operational tasking. In the summer of 1956, the Air Force Reserve began taking on flying missions which had always been the domain of the active duty Air Force. Troop carrier wings brought U.S. Coast Guard equipment from New York to the Caribbean. The next spring, AFRES aircraft flew airlift missions for the Tactical Air Command. The next year, equipment was transported by AFRES aircraft between Air Material Command bases, aeromedical units were flying mercy missions, and the troop carrier wings were dropping U.S. Army paratroopers on a regular basis. Pentagon planners began to glimpse the expertise available in the Air Force Reserve and the cost effectiveness of AFRES operations.

Eisenhower Administration cutbacks in 1957 caused a drastic slashing of the Air Force Reserve to 45 troop carrier squadrons and five air rescue squadrons. The fighter mission was moved to the Air National Guard.

Cold War mania during the Eisenhower years also caused Pentagon planners to view the Air Force Reserve as a source of post-strike and recovery manpower in the event of a nuclear war. Until the concept was abolished in 1965, many air reserve centers were established for classroom training in support unit duties.

The first post-Korea mobilization test came in 1961 during the "Berlin Crisis" when President Kennedy activated two troop carrier wings consisting of five C-124 *Globemaster II* squadrons to augment the airlift force. The 435th Troop Carrier Wing from Homestead AFB, Fla., and the 442nd Troop Carrier Wing from Richards-Gebaur AFB, Mo., were activated on Oct. 1, 1961, and released on Aug. 27, 1962.

Two months later, President Kennedy again turned to the Air Force Reserve during the "Cuban Missile Crisis" and on Oct. 28, 1962, activated 14,000 reservists in seven C-119 and one C-123 troop carrier wings, plus six aerial port squadrons for the month of the crisis.

The major operational change for the Air Force Reserve came during airlift missions to Vietnam. Although no AFRES units were called to active duty until the Tet Offensive of 1968, the seeds of two major Reserve concepts were sown in the need to fly troops and supplies across the Pacific to Vietnam.

STRATEGIC AIR COMMAND

Cold War fears bring SAC's nuclear bombers to Westover

The Strategic Air Command officially took control over Westover AFB on April 1, 1955, with the activation of the 4050th Air Refueling Wing of KC-97 *Stratotankers*.

There was massive construction at Westover in the mid-1950s. Some was visible, such as the 300-foot wide main runway that stretched for 11,600 feet. The first B-52s arrived at Westover in the autumn of 1956. More hangars were built to repair B-52s and nosedocks for KC-135s.

But there were also secret construction projects undertaken between 1953 and 1954 which were never visible to the residents of Western Massachusetts or most employees and service members on the base.

Along the eastern portion of the base by Ludlow, a highly classified facility was built within an electric three-ring fence to store and repair nuclear weapons. Known as Stony Brook Air Force Station, the facility was a separate installation within the installation of Westover.

Another classified construction project involved digging deep into Mount Holyoke for a nuclear bomb-proof alternate command post for the Strategic Air Command. Known by the local site nickname of "The Notch," this facility began opertions in 1956 and was later a key installation during the Cuban missile crisis.

A third secret project was known as building P-1900, a large photographic laboratory and mineral recovery center where reconnaissance photography was developed, analyzed, and turned into secret maps of tremendous detail. The huge "Air Force Special Projects Facility" was self-sufficient with billeting, food and medical areas along with the classified work sections.

Tribute can be paid today to the men and women of Westover AFB who kept these secrets securely during the Cold War. There is no public record of security breaches, no FBI indictments or historical literature detailing any national security compromise coming out of Westover AFB during the era when massive retaliation by B-52s carrying nuclear bombs seemed to assure peaceful coexistence.

The Strategic Air Command was created Mar. 21, 1946, to perpetuate the strategic bombing capability of the U.S. that was so painstakingly developed during World War II out of the vision of Oscar Westover and other tacticians. The decision to base SAC and the mighty Eighth Air Force at Westover AFB was an unplanned tribute to the general who had criss-crossed the nation talking about "a flying fortresses type of bomber" and pressured the War Department into giving the B-17 a chance.

One of the first efforts by the new command after World War II was to rely on nuclear weapons and to pare down America's conventional bomber force. At the end of the war, the U.S. owned 3,063 light bombers; 5,576 medium bombers; 12,883 heavy bombers -- including 5,267 B-24s -- and 2,242 very heavy bombers, as the B- 29s were designated. Within a year the force was down to only 15 B-24s, a handful of B-17s and 2,112 B-29s.

A NUCLEAR MONOPOLY and more than 2,000 B-29s gave confidence to the new U.S. Air Force in the postwar years as Strategic Air Command planners awaited the massive B-36 Peacemaker.

In development when the Air Force was formed in 1947 was a pre-war idea, the "hemispheric bomber," a long-range aircraft so massive that when it entered the inventory the so-called very heavy bombers -- the $315,000 B-29s -- would be redesignated medium bombers. The Convair YB-36A began test flights in August 1947 and the project was the largest item in the service's budget.

Initially ordered by the Army Air Corps in 1941 as an inter-continental bomber that could bomb Berlin from East Coast bases like Westover Field if England fell to the Germans, the huge aircraft weighed more than six times as much as a B-17 and could carry a 10,000-pound bomb load 10,000 miles.

The first of 385 operational B-36 *Peacemakers* was delivered to SAC on June 26, 1948, and the aircraft became the symbol of the new Air Force and America's strategic might as a nuclear power.

Under General Curtis E. LeMay, SAC developed a network of B-36 air bases ringing the Soviet Union with nuclear bombs.

The huge bomber also became a symbol of Air Force controversy due to three events occuring in rapid succession; two caused by the Soviet Union and one within the procurement process. Capitalizing on captured German technology, the Soviets unveiled the MiG-15 jet fighter which made the B-36 vulnerable. Defense Secretary Louis Johnson had served on Convair's board of directors during the development of the

B-36s arrive at Westover AFB

Carol Bernier-Gawle remembers the dawn of the SAC era when she was a ninth grade student and had made friends with a Westover dependent named Betsy. "She was the first person I had ever met with a Southern accent. And I think that was the beginning of my education as a cosmopolitan person rather than just confined to the city of Chicopee. For some reason we started getting a lot of different people from a lot of different places who had been to a lot of places."

Betsy had been to Turkey. "Up until then, Turkey was just some place in geography on a map. But here was a person who had been to Turkey and had lived there." Many of the new students in local schools brought a worldliness to the area.

"Some of the kids I met had been born in England, or France, or Africa, because their parents had been stationed there in the Air Force. All of a sudden, all these people were coming in. We started hearing things like 'Strategic Air Command' and we started seeing a lot of different planes flying over. And one of them was one of the most beautiful planes....the B-36. It was the most impressive, huge plane I've ever seen," she said.

"I remember the first time one flew over. The whole house rumbled, it shook, it vibrated. There was this deep vibration and I remember saying, 'What was that?' and I remember running out on the porch, looking up, and seeing this great big bird of silver. Beautiful....but they didn't last very long there (at Westover). They were here a short time and then the B-52s came in."

aircraft and this led to Congressional hearings on the propriety of the Air Force procurement process.

The third event caused a significant turn in American policy. In August 1949, Air Force aircraft detected radiation in the atmosphere over Asia, indicating the Soviet Union had detonated a nuclear device. With the expensive B-36 now termed "a sitting duck" and the U.S. having lost its nuclear monopoly, the Strategic Air Command took on new importance and there was a drive for major construction at bases, such as Westover, to support SAC.

EMERGENCY LANDING at Westover in 1949 by this B-36 from Carswell AFB caused extensive damage to the right wing and an engine when the aircraft struck a radar shack. The B-36 wing was repaired at Westover and the aircraft made a five-engine take-off.

One way around the MiG-15 interceptor problem was to develop a high-speed medium bomber and to produce the aircraft in great quantity as insurance that nuclear weapons would be delivered inside the Soviet Union. The swept-wing Boeing B-47 *Stratojet* was first delivered to SAC in 1951. With a top speed of 606 mph, the B-47 was much faster and more maneuverable than the 381-mph B-36. By the end of 1953 there were 329 B-47s operational with SAC and the number eventually rose to more than 2,000. The last B-36 was retired in 1958.

PRESIDENT Eisenhower was faced with the Soviet capability to deliver atomic and hydrogen bombs to North America. To counter the threat, he called upon General Curtis LeMay and the Strategic Air Command to provide around-the-clock retaliation.

The most famous bomber associated with Westover AFB was the B-52 *Stratofortress*. While the B-47 was in production, Boeing designers were developing an aircraft nearly as large as the B-36, but with the speed of the three-man B-47.

First flown in 1952, the first of 639 B-52s was delivered to SAC in June 1955 and the first B-52 arrived at Westover AFB in September 1956. The new aircraft were assigned to the 346th and 348th Bombardment Squadrons. The units had been flying B-36s at Fairchild AFB, Wash., as the 99th Strategic Reconnaissance Wing (Heavy), but were transferred to Westover to take delivery of the new $8-million B-52s as squadrons of the 99th Bombardment Wing.

The original 99th Bombardment Group had a distinguished World War II record from operations out of North Africa. In 395 missions, 99th Bomb Group crews flew more than 10,000 sorties in B-17s to drop more than 24,000 tons of bombs. Throughout 1943 the 99th flew missions over Italy, Germany, southern France and Greece. In February and March 1944 the 99th gained a reputation for bombing accuracy over Cassino,

BRAND NEW B-52s began arriving at Westover in 1956 for the 99th Bombardment Wing (Heavy). The undersides of the Stratofortresses were painted white to reflect the energy from the searing fireball of the nuclear weapons they were designed to drop.

Verona and Turin, Italy. During the remainder of the war, the 99th bombed railways and oil refineries, including the famous raids on the Ploesti oil fields in Romania.

To provide in-flight refueling for the B-52s, the Air Force purchased modified Boeing 707s designated as KC-135 tankers. The 707 first flew July 15, 1954, and the first KC-135s were delivered in 1956 and arrived at Westover with the 499th Air Refueling Wing, which was redesignated the 99th Air Refueling Squadron and assigned to the 99th Bombardment Wing.

A modern $3.5 million USAF regional hospital opened in 1956 with a staff of nearly 400 physicians, dentists, nurses, technicians, corpsmen and administrators. There were three in-patient wards for dependents, surgical and medical patients; a dental clinic, medical laboratory, X-ray department, pharmacy and physical therapy department.

The hospital had an impact on Greater Springfield. Air Force doctors moonlighted at local hospitals and some physicians remained in the Pioneer Valley. One group of physicians completed their military service and established an emergency room service firm which contracted with many local hospitals to provide trauma treatment.

In the 1960s, Westover's 57th Air Division was one of the largest organizations of its kind in the Air Force. The unit had been activated at Westover on Sept. 4, 1956, and was responsible for sustained combat readiness of B-52s and KC-135s while operating under a condition of continuous alert. During the Cuban Missile crisis, the 57th Air Division demonstrated just how efficient the unit could be at Westover. In 1963, the 5,000-man 99th Bomb Wing won the Selser Memorial Award as the outstanding bomb wing of the Eighth Air Force.

The 57th also controlled the 68th Bomb Wing at Seymour Johnson AFB in North Carolina, and Westover's 99th Bomb Wing and 499th Air Refueling Wing. The 499th ARW's motto was *Auxilium ab Alto* -- "help from above."

Included in the 57th Air Division was the 814th Medical Group and the 814th Combat Support Group at Westover which maintained more than 600 buildings, 72 miles of roadway and 20 hangars in 1964. By 1972, the unit maintained more than 1,000 buildings, 88 miles of roads and 32 hangars.

Although each unit of the division was based within the U.S., the range and mission of the three wings were global. The B-52 *Stratofortresses* provided a force designed for immediate and long-range offensive bombing with nuclear or conventional weapons. Aerial refueling of the B-52s was the job of the KC-97s and KC-135s.

***AIR POLICEMAN** stands behind concertina wire near Runway 33 to guard KC-97s of the 384th Air Refueling Squadron. In photo below, SAC crewmembers of the 384th AREFS mark the squadron's last flight and note their 1,000 consecutive on-time takeoffs.*

Fifteen die in KC-135 crash

Kaz Machowski was 12 on June 27, 1958, when a string of KC-135s took off from Westover's Runway 23 bound for a record-setting mission to Europe. "Most vivid in my mind is getting up out of bed, looking out the rear window of the second floor. That's where the bedroom was, looking out through the venetian blinds and just seeing flames and wreckage."

The tanker, with 15 crew and journalists aboard had struck the Massachusetts Turnpike and crashed into the farmland behind Machowski's home.

"I knew it was an airplane," he said. "Unbelievable, fire, flames and wreckage. I remember seeing the tail section of the plane in the wreckage, upside down, with the tail pointing down in the field. Seeing part of the landing gear about 15 feet away from the back of the house, ablaze.

"The next thing I seem to recall was the large Air Force fire trucks. Big lumbering monsters coming down Fuller Road. And at that time Fuller Road was a dirt road, pretty much and they had to come up an incline to drive into our cornfield. And I can just remember standing there just watching this big red fire truck just growling right over the embankment. And driving through the corn, because that's what was planted in the field. Just going off and fighting the fire. Air Force personnel -- after the fire had been pretty much under control -- of them walking through with body bags."

The next morning everything was still smoldering, he said. "The police of course still there, the fire trucks. Air Force personnel had been set-up as guards. They started roping off the area."

KC-135 tanker departs Westover's Runway 23.

Westover's 337th Fighter Interceptor Squadron

On Aug. 18, 1955, the 337th Fighter Interceptor Squadron was assigned -- on paper, no personnel were transferred -- to Westover AFB to replace the 60th Fighter Interceptor Squadron. In the spring of 1957, the 337th switched from F-86Ds to F-86Ls and the squadron became a unit of Westover's 4729th Air Defense Group.

The 337th had an odd history before coming to Westover. Unlike most units that were created in the U.S. and sent overseas, the 337th was activated in Iceland on Sept. 11, 1942. The initial mission of the squadron was to deter a Nazi invasion of Iceland, which was a strategic spot to protect American shipping bound for the Soviet Union and the United Kingdom.

But before all the personnel and equipment arrived in Iceland, the 337th FIS was reasigned to Glendale Army Base in California as a Lockheed P-38 *Lightning* unit. Personnel for the 337th came from the 329th Fighter Group headquarters and the 330th, 331st and 332nd FIS.

During the early stages of the war, the 337th FIS was one of five fighter squadrons assigned to defend the West Coast. As the war tilted toward offensive operations in the Pacific, the 337th became a fighter pilot training unit at Muroc and Ontario Army Air Fields in California and Paine Field in Washington. On Mar. 31, 1944, the squadron was absorbed by the 443rd Army Air Force Base Unit at Ontario Field.

As the Korean War was ending, the unit was given new life as the 337th Fighter Interceptor Squadron and was activated July 8, 1953, at Minneapolis-St. Paul International Airport to replace the 18th FIS that was reassigned to Alaska. In Minnesota, the 337th flew F-86D *Saberjets*.

After coming to Westover in 1955, the 337th FIS moved from F-86Ls to F-104A *Starfighters* and was assigned the mission of air de fense of the Northeast for the Eastern Air Defense Force. In the early 1960s the 325-man squadron converted to F-106 *Delta Dagger*. With the air defense mission, the 337th was required at Westover to have pilots and aircraft on alert 24 hours a day.

F-106 of the 337th FIS at Westover

Harry Wolfe, a former intelligence officer described Westover in the early 1960s as a "fairly busy place. In the days when SAC was very active with constant training. At that time, fly, fly, fly was the name of the game. Being in SAC in those days was like being inside a pencil sharpener. You constantly had a sharp edge, but you were being worn down because it was very demanding physically."

There were early morning take-offs. "You would fly for eight to 12 hours depending on the particular demands of the mission. Frequently there were in-flight refuelings which extended your time. And then, of course, you'd get down and land at most unusual hours. And maybe you'd have half the morning off the next day, but you were pre-flighting again for the next mission."

He said part of the weekly routine in the intelligence section was to spend time refining individual crew war plans. "Designing missions to simulate the conditions you have run into. The types of navigation. For example, if your plans had you flying over the Northern areas, above 70 degrees north latitude, there was a different navigational technique that was used called polar navigation. Of course, your magnetic compasses up in the northern latitudes are no good. They tend to dip straight down. So there was an entirely different system that was used. In addition we polished our celestial navigation," he said.

"We were constantly practicing bomb runs on the major areas around the country. The bomb runs were made and tracked by a radar system and it was just go, go, go!" Wolfe said.

KC-135 of the 99th Air Refueling Squadron taxis onto the north ramp where Air Force Reserve C-124 Globemaster IIs are parked. Nosedocks on the right were used for KC-135s.

The Red Barn and Ma Manning's

The Red Barn and Ma Manning's were landmarks during the Forties, Fifties and Sixties. "That's a sentimental journey for plenty of GI people," said Neil O'Leary. "When a visitor pops into my store, reaches over and says, "Whatever happened to the Red Barn?" That's a common question. Because it was a great place for the single fellow away from home. They had good food, entertainment, a large bar, good parking lot, everything you could ask for, it was so close to the base. Everything you could ask for."

The Red Barn was where Medical West is today, said Betty O'Connell, a retired Westover civilian. She remembers the start of the legend when she was working for SAC in the mid-1950s. "We were working shifts and we used to go out after our shift ended about 11 or 11:30 at night. We'd go over to the Red Barn. We'd ride horses. One night the owner said to us, 'How would you like it if I put a jukebox in here and got you a keg of beer?' and we said, 'Hey, that would be great.' And that's

LATE FORTIES pen and colored pencil cartoon, signed by Fejes, which was displayed behind the bar at Ma Manning's. *(Courtesy of her niece, Lila Hanh of Chicopee.)*

how the Red Barn started and became a mecca for anyone coming in and out of Westover."

"It was an old barn, a horse barn," said O'Leary. "The Sullivans or somebody like that had horses in there. And there was a silo fashioned into the kitchen. The bar was an addition, a huge two-level bar. The main dining room had a level or two. They had some entertainment there, Wednesday through Sunday. The usual -- the comedian, the dancer, or the singer. It was a great place for the GI to get off base and spend a few dollars and forget his loneliness."

O'Leary said, "Ma Manning's was somewhat the same, but much smaller. She had a wall of memorabilia that was second to none, and those guys contributed to that from day one. Guess that bar was there before Westover even began, but it put the bar on the international scale. It was known everywhere. Great lady. She cashed checks, took watches, helped guys out, the usual stories that are around military bases. A fellow could go in and cash his check at night if he didn't get it done during banking hours."

According to O'Connell, "You could travel around the country and the world today and talk to people who still knew the Red Barn and Ma Manning's. And many's the good party we had there and at Ma Manning's.

"Ma Manning's, I swear there was never a place in the United States like Ma Manning's," said O'Connell. "She loved all the guys. She was good to them and I doubt very much if she ever got stuck very often. She would just want to make you feel -- and she was a tough old cookie -- but she'd want to make you feel at home there. We used to go there on Friday nights and get all the clams you wanted for 25 cents, drink beer and play the bowling machines. It was never fancy, even when we had our parties, but we all managed to have a good time there."

Red-haired "Ma" Manning was actually Josephine Hess, born in Denmark in 1890, who moved to Chicopee as a child and later married Manning Ross, who opened Manning's Cafe by James and Prospect streets in 1936. After he died in 1941, his 51-year-old widow was in debt, but Westover Field was beginning to blossom. "She poured a lot of love into those lonely GIs," said one former patron.

MA MANNING, the Danish-born Josephine Hess who married Manning Ross, behind the bar at Hangar 13 with an Army Air Forces customer.

HANGAR 13

She held Saturday night dance contests and had a ballfield for Friday night double headers with beer and steamed clams, and Sunday softball games and picnics.

Stories are told today about how "Ma Manning" would spot airmen $10 in exchange for their watch, and on payday she would be behind the bar wearing an arms-worth of wristwatches.

She bought a new Cadillac each year and opened mail from military bases all over the world. When she would tour Europe or visit other states, former Westover airmen would greet her with parties and at least once with a band. A nephew serving overseas once reported seeing Ma Manning's name and an affectionate note scribbled on the wall of a Japanese latrine. She sold the bar in the Seventies and died of a stroke, at age 90, in February 1980.

Like Manning's Cafe, the Red Barn earned a wide following. Once the Red Barn became world famous, O'Connell remembers, there was a change in the off-base watering hole. "There was just some thing about the whole atmosphere there that we all loved to have our parties there and they had a lot of big names come in there for entertainment. It never went back to the way it was when we first started going out there when it really was a barn. I was talking to the owner one day and he said the operation was getting too big. He said, 'Now I worry every week if I'm going to be able to meet my payroll.' They even had their own bakery there by that time. It was a big operation.

"I don't think anybody who ever lived in Chicopee hasn't been to the Red Barn, sure. But we used to use their grove a lot. They had some tremendous steak roasts out there. They had everything. That was quite a big facility," O'Leary said.

"I can remember the 16-ounce steaks better than anything," he said. "The Pappas Brothers ran a tremendous kitchen at the Red Barn and if you wanted a super-sized steak, that was the place to go. They were great food men. They still are. Go over to the annual Glendi festival in Wilbraham and you can see the Pappas brothers cooking steak on the shishkabob. Their steak was second to none, my best Red Barn memory."

In 1988 the Soviet Union admitted the existence of a thoughtful and thorough 55-year-old program to confuse outsiders by publishing inaccurate maps. This was known in Soviet jargon as a *"maskirovka"* program, one intended to deceive an enemy. The announcement was no surprise to the U.S. intelligence community. Spy satellites coded with names like "Big Bird" had for years been dropping film to specially equipped C-130s to be analyzed. In later years other orbiting platforms, like the "Key Hole" series KH-10s, had been transmitting data to ground stations. American cartographers knew for decades that the pictures didn't match the Soviet maps.

But military cartography has always been a secret art, and the early link of Westover AFB with clandestine map-making was highly classified. Thousands of maps were made at Westover during the SAC era, and these maps were created from photographs taken by secret airplanes.

Betty O'Connell remembers little red and white aircraft bringing secret film to Westover. Other people remember twin-engined blue and white planes under contract to the Air Force flying to Westover so men could deliver leather pouches to the 8th Reconnaissance Technical Squadron in building P-1900, near where the James Street gate is today.

General Curtis Emerson LeMay

Although the nation and many people of the world would be shocked on May Day 1960 when a high-flying Lockheed U-2 was downed over the Soviet Union and pilot Francis Gary Powers parachuted to safety near Sverdlovsk, few people outside Westover ever put together the whole puzzle.

Information equates to power in most institutions, but even more so in the military where there has been a longstanding belief that the agency which controls reconnaissance, determines the threat, shapes the response, and becomes a powerful empire within the institution. As SAC commander, General Curtis E. LeMay wanted to control American spy plane reconnaissance. He wanted SAC photogammetrists to analyze the film and determine the Soviet threat to

America, and he wanted SAC to be the primary response to any global threat. To control the information and to make the target maps in-house at Westover was to assure that what SAC needed, SAC got. And once SAC received U-2s in 1957 and assigned them to "weather squadrons," LeMay's empire got quite a lot at Westover.

"Recce Tech" was the nickname of the 8th Reconnaissance Technical Squadron, and the multi-million dollar building P-1900 was known on base as the "Air Force Special Projects Facility."

"It was always hush-hush," remembers Betty O'Connell. "We were told it was like a city within a city. Once they went in, they didn't come out all day. They had a beautiful cafeteria, they had everything self-contained."

Harry Wolfe described the activities inside P-1900. "Very much involved with transferring aerial photography into map-making. And also map production and distribution. When I was assigned to Westover, I was assigned to Eighth Air Force headquarters for duty, but for morning report purposes my headquarters was 8th Reconnaissance Technical Squadron. I couldn't tell you who the commanding officer was, because I never saw him," said Wolfe.

The 8th RTS operated a phenomenal photographic laboratory in the special projects building. "I'm sure you've seen the building," said Wolfe. "Very busy. And very classified and with limited access. There were two sections to that area. One was a basic photo lab. Tremendous equipment. With it was a map-making facility where they drew the plates for creating new maps."

Because maps were made of potential target areas -- even maps of areas which U.S. politicians would be forced to defend in public as "allied nations" -- the cartography was highly classified. The source of the cartographic information -- U-2 spy planes in the 1950s -- also, of course, forced a heavy veil of secrecy around the Westover project.

The piercing sharpness of the Hycon Model 73B aerial camera aboard the 500-mph U-2 that could fly nearly supersonic at altitudes above 70,000 feet was one of the top American secrets of the Cold War. And most Westover veterans believe the secret was kept for years at the base.

"Whatever they did was supposed to be so hush-hush," said O'Connell, "that if you went to the bathroom you had to have somebody go with you."

She estimated 10 men worked daily to keep the air conditioning running properly in the huge facility where fumes from processing chemicals and a silver recovery system had to be controled. "They had their own medical staff there. Once you came in during the day you didn't have to go out for anything."

The Cuban Missile Crisis
October 1962

Harry Wolfe, a former navigator and target intelligence officer with Eighth Air Force remembers well the secret start of the Cuban missile crisis at Westover. During his time at the base the Eighth Air Force headquarters was in the old hospital complex fairly close to the main gate. Across the street was the Target Intelligence Section surrounded by barbed wire, "a caged-in building that had been the old venereal disease building in the 1940s. But it was good for us because we had good security and the high wire fence," he said.

"The first indication I had that something was up was I heard whispering in the back office. Then I was called in by my boss, and he said, 'You're going to be navigator on an aircraft that's going to fly to Rochester. We're taking off tonight. Go home and pack a bag!'

"THE NOTCH," 400 feet below the summit of Mount Holyoke, was the alternate headquarters for the Strategic Air Command.

"This was early afternoon, I believe. So I did as ordered. I went home and got a little bag, a shaving kit, and came back out to Westover," he said.

"We were flying a T-29, a twin-engine, reciprocating engine. And there was a general officer aboard. And I was told we were ferrying some film to Rochester, New York, to be processed -- to receive special handling and processing. And I was told, unofficially, it was U-2 film. And it was very, very important," Wolfe said.

AERIAL RECONNAISSANCE photographs proved the Soviets were supplying intermediate range ballistic missiles to Cuba.

"I was familiar with U-2 photography from the past. So when they said U-2 film, I said, 'Okay...' We flew it there. The general officer aboard was in charge of the film and I thought that was a little unusual. And I had suspicions. I heard the word 'Cuban' and that sort of thing in our little area. Very thin walls and everybody was cleared top secret. So it really wasn't violation of security. We flew the film to Rochester. We dropped it off, and then we came back the next day. And when I got back it was confirmed that yes, indeed, this was U-2 photography of Cuba and they believed it was going to show ample evidence of missiles in Cuba."

SAC HEADQUARTERS monitored the movement of Soviet vessels bound for Cuba.

Wolf said Westover and the Eighth Air Force, which was responsible for the eastern third of the nation, went into a condition of alert. "Which mean't that intelligence people reported to the Notch, the site out in the mountain -- the Eighth Air Force auxiliary headquarters," he said.

"My particular job was monitoring the Russian ships. There was one, I think it was the *Bertorri,* which supposedly at that time was carrying some additional missiles. And it was this one that was the key to whether or not Khruschev was going to withdraw his missiles.

"We had aircraft observations...we were getting immediate reports from aircraft. We had constant reconnaissance coverage of all the vessels in that area," Wolfe said.

"After several days of this -- and the intensity was picking up, and I think some of it was starting to show in the newspapers -- we began to wonder when we went to work in the morning whether or not we would have a residence to come back to, because it was getting real serious. In fact, it reached the point of seriousness where the SAC commander -- and I think it was Tommy Powers at that time -- made an all-command broadcast to the command posts.

"As I listened to Powers' speech. The thought that went through my mind was George Washington's farewell to the troops. Of course, Washington said that at the end of hostilities. But it sounded like the general was giving his farewell at what might have been the beginning of hostilities," Wolfe said.

"I think I was just in awe that we had reached that point where the SAC commander would essentially say, 'We hope to see you later.' That's not what he said, but it came out that way."

Wolfe spent most of this crisis period in "the Notch" some 400 feet below the summit of Mount Holyoke. "We did have this tremendous mountain sitting over us. We all had concerns about our families, but it didn't show itself in our work....Some of the officers I worked with did send their families away from Westover in the possibility if something did materialize out of this. I had considered doing the same thing because I did have family in Western Massachusetts....I debated with myself whether to move the family out or not. The one thing I was concerned about was, 'If I packed up my family and left, what are the neighbors going to think? What's going to happen? What kind of an atmosphere will that create?'

"By then of course, the news was in the newspapers and was coming over the TV and everything else. And people knew the situation was serious. But if my family just packed up and took off...I was very concerned about the overall effect," he said.

Michael Daly was a SAC navigator assigned to the 99th Bombardment Wing at Westover AFB from 1964 to 1968.

"I was just impressed with Westover and the vastness of the mission we had within the Strategic Air Command," said Michael Daly, president of Baystate Medical Center, a former Air Force captain and B-52 navigator who arrived at the base in 1964. "Westover was reknown in the Air Force for having probably one of the highest quality runways in size, length and capability of any in the United States and it was premiere facility for the entire Air Force. Among air bases, Westover was just held in the high est of regard by the pro's who knew about flying."

Daly had grown up near Hartford and was graduated in 1963 from Trinity College. He entered the Air Force in 1962 and spent two years at training bases before arriving at Westover AFB for his first active-duty assignment.

Daly was impressed that Westover "was a headquarters air base for the Eighth Air Force and had so many functions attached to it. The ancillary support system suggested much more capability than one would encounter at a normal base. There was a certain amount of confidence associated with that," he said.

"I know the attitude I encountered was very positive, but the Strategic Air Command bases, generally, were the elite of the air crews and you would find high attitudes virtually anywhere you went. I can't say that our morale or attitudes were better or worse than any, but certainly very high," said Daly.

He described the lifestyle of SAC aircrews at Westover during the mid-Sixties. "Our schedule as a bomber crew was such that we didn't have a typical day. We ran in cycles of about three weeks duration where you would have one full week on alert -- which meant for seven days, 24 hours a day you lived in the Molehole with your crew.

"That was the highest stage of readiness and you were prepared within a moments notice to get to your aircraft, get your air craft up and going and ready for launch within a couple of minutes. That required a certain amount of intensity and alert ness. During that one-week period, your adrenalin was running as high as ever. When you got off alert -- which was a very formal activity, being on alert, reporting in, getting off -- you got some days off, and then you went into a two-week period when

MICHAEL J. DALY is today president and CEO of Baystate Health Systems Inc., the parent corporation of Baystate Medical Center in Springfield and Franklin Medical Center in Greenfield, Mass.

B-52 STRATO-FORTRESS

you were flying training missions and honing your skills as a crew," he said.

"A lot of time was spent on the ground studying various missions your crew would fly, different types of tactics you would employ during those missions and getting your skill levels increased. You actually never really flew on alert. On alert you prepared to fly and when the claxon went off, you jumped to your aircraft and you were either going on one of two assignments -- you were either being launched because a situation had evolved inter- nationally that required us to be airborne, or, secondly, about once a year, they would have an Operational Readiness Inspection and the Inspector General would show up and we would launch from alert on test exercises. All of our in-flight training, with the exception of those two instances, was done in the period when you were off alert. That was a time of study training and flying," said Daly.

"The planes in part of the aircraft associate with the alert crew. There were another 12 to 14 bombers immediately adjacent, down the runway, which were also on alert. We would have about 15 to 20 crews on alert so it was just not the Christmas Tree, it was the Christmas Tree plus. We're talking six guys to a plane, the six alert planes and up to 20 other planes, that would be a lot of people. And also on alert were the maintenance crews who had to respond in the same way that we did in getting the aircraft ready within seconds. There were aircraft maintenance personnel who were also in housing that allowed them to get immediately to that aircraft within a very short prescribed time.

There was another housing complex about 30 seconds away from the Molehole. All crews on alert ate their meals in the Molehole, but not all were housed there overnight. All of the prescribed study that you had to accomplish while on alert occurred in that setting," he said.

"There were new techniques for bombing runs, studying how you flew low-altitude versus high-altitude missions. Various members of the crew would be introduced to new types of equipment being brought into their work stations. We would all go through briefings and training sessions. The Molehole was the center...the only place where all activities associated with alert were housed," said Daly.

"I remember the major entrance from the parking lot was used the most because literally every day on alert -- when we weren't involved in study or scheduled activities -- most of the guys met their families -- you could see your family at least once a day. We would all walk to the parking area. I remember that particular tunnel was the best. When you entered through the parking lot through that tunnel on your left I can remember the cafeteria. Beyond that, on the left was the main briefing room, enter tainment area. It was the single largest room in the complex. Also on that floor were three fair-sized conference rooms where you would break off either with your crew or people who were in your discipline, in my case a navigator and do various things you had to accomplish over the course of your seven days on alert. The lower floor tended to be principally a barracks type, room type set up," he said.

"We would sleep in the same building, travel in the same vehicle. If one of our colleagues had to travel to a particular building on the base, we all had to go. It used to be allowed that we could go to the BX to do some shopping. That was in the perimeter of distance we were allowed to travel. If one of the guys wanted to meet his wife, we'd all go. If somebody had an annual physical, we'd all go to that. They tried to schedule us much together as a crew as they could, but there were times when six guys would travel to allow one to conduct some business. It wasn't a problem. You did get to know one another as a family. I really enjoyed my crew and enjoyed being with them. We could not have survived without the high degree of mutual respect," said Daly.

"There were transfers for reasons. I'm sure there must of been personality clashes, but I never encountered that. I think for the four years I spent at Westover, I spent 75 percent of my time with a single crew, and I went to Vietnam with that crew," he said.

"Life in SAC could be tedious at times, but we were in an era of the Cold War. There had been a confrontation not too many years before between the United States and Russia. We believed -- and we believed sincerely -- we were on the front line of our nation's power to deter a nuclear war. That was a heavy responsibility. I don't think that was ever out of our minds. Ever. When we were participating in any exercise, you might get bored with the routine nature encountered over the course of several days, but the minute the horn went, all boredom ceased. There were enough practice exercises that you were off-and-running and the adrenalin was flowing and you didn't know if this particular exercise was the real thing or you had another practice. There was a

THE MOLEHOLE where SAC alert crews were on duty 24 hours a day.

B-52 ROARS down Westover's main runway at dawn.

certain amount of boredom, but we were honed to a pretty fine degree and felt that we could be launched at a moment's notice."

Daly said SAC crewmembers were also fully aware of the potential of the nuclear weapons carried in the bomb bays of Westover's B-52s. "The frail nature of international relations -- when you have to employ that kind of deterrent, that kind of a tactic -- was significant. At the time, we were absolutely convinced what we were doing was in the finest interest of this country and we were in a role that would prevent nuclear warfare simply because we were prepared to respond. We were never ever prepared to be the initiator. That was never part of our role. I think just that high level of capability of response was something we in SAC were one, proud of, and two, frightened of," he said.

"I think every military person, by the time you reach that stage of assignment which was about as high and important of level of assignment that you could receive in the military to be trusted with that kind of a role. You'd gone through several years of schooling, study. Not only in the task you had been assigned, but in the affairs of the world and the motives of the United States. We felt that we knew what was occurring. We did have great confidence in our leaders. Back in the early Sixties the United States was revered around the world as the country promoting peace around the world. That was our motto, 'Peace is our Profession,' and we were proud of it."

Daly said people residing in Greater Springfield were well aware of the roaring B-52s and KC-135s. And they were also aware of the nuclear weapons stored on the base. People may not have known the weapons were assembled at Stony Brook Air Force Station behind triple electrified fences, but they knew atomic bombs were at Westover. After all, they might say, what are the B-52s here for anyway?

"The community was aware of it as the most sophisticated weapon available to the United States at that time. Again, the media would point to Westover and it's capabilities with pride. We were part of the community. Certainly we would stay together as military types, but a number of people like myself, lived in the community not on the base. I never encountered anything but healthy respect for what we were doing on behalf of the citizens of this region and the United States," said Daly.

"I think as young as some of us were at the time, when I was in my early 20s, I was a college graduate, was aware of what was transpiring in recent years in international relations. I was aware that the capability of the Strategic Air Command was a critical ingredient to the power of the negotiating position we had as a country. I was very much aware that we were perhaps playing a key role in preserving world peace," he said.

"I encountered a great deal of pride for the level of capability that was presented through us and Westover Air Force Base. Also, I recall quite vividly the types of newscasts that were prevalent during that period. The very popular movies of the day were movies that were very positive in terms of the capability of the Strategic Air Command had developed. There was a high level of awareness certainly not to the degree of the aircrews themselves, but I felt that people were very much aware of what we were doing....We were critically aware of the magnitude of the equipment and the dollars spent, weapon systems that were developed, that six of us might be called upon to employ. It was a very significant responsibility that we felt," he said.

But in Captain Daly's years at Westover, none of the crews were ever launched amid looming possibilities of hostilities. "But when the horn sounded, we went every time as though we had been launched in some sort of a warfare situation. It was only at the point of actual taxiing and takeoff that it was declared a practice exercise. We would go to the end of the runway prepared to launch and then they would broadcast: 'This has been an exercise, please recover your aircraft to the designated area.' But whenever that claxton would go off, we would say things like, *Good grief! This might be it.* Clearly that was the reaction. What ever you were doing, you would bolt up, look at your partners or crew members and you would be off-and-running. You had been so finely trained that you didn't have time to wonder, to doubt because there was so much to be done from the minute that horn sounded, you just had to go, almost by because there was no time to wonder, doubt, worry what might be out there. That happened after the fact. You had to get in that plane and get it ready to go," he said.

"It was only at the point of actual taxiing and takeoff that it was declared a practice exercise."

"If you actually got launched and were assigned a target in a hostile country, you might launch and expect several points during the mission as you proceeded for that target to pass various points where in fact you would receive a message from command that the mission was still on or has been cancelled. I really can't recall, but there was probably about two of those points. At the final point, if you were to reach that, you were pretty much relegated to know that a nuclear holocaust had occurred. None of us wanted that to happen, but at that point we had been lead to believe that the hostilities had already commenced and that we had no alternative but to retaliate," said Daly.

"I was terribly disappointed to see that capability removed from Westover. It was a key ingredient of the life of this region to have that kind of military capability available as a community resource, a point of pride. A sense of disappointment when it was lost" in the early 1970s.

Daly said his life changed from the routine of alert and training when his B-52 crew was assigned in 1967 to bomb South Vietnam as part of Operation Arc Light. His wife was a registered nurse working at Wesson Memorial Hospital in Springfield. She took their son to live at Daly's parents' home in Hartford.

"Westover's first involvement was in '67....We would go over for approximately seven months -- and we were expected to fly as many as a hundred combat missions. When my crew deployed in September 1967 we took about 16 to 18 hours from Westover get to Guam, which was our initial station. Although during the course of our Arc Light deployment, we found ourselves based not only in Guam at Anderson AFB, but from time to time we would find ourselves in Okinawa and my crew was one of the first crews to open up the base in Thailand when we were deployed there," he said.

"When we launched, the Vietnam War was at its peak, its most massive involvement. It was late in '67, early in '68 that we reached the Tet Offensive which historians mark as the high point of the war in terms of commitment and involvement of the most troops, the most dollars, the highest stakes....I think in the post-Tet period there was a totally different environment and attitude as the United States started to withdraw. I didn't see that. I was involved in that whole period where the United States appeared to be more and more committed to that engagement. There are many chapters in the Vietnam War. I hap-

99th AVIONICS MAINTENANCE SQUADRON technicians repaired the sophisticated electronics required by the mission of the 99th Bomb Wing.

pen to be in the one that is perhaps similar to the United States commitment to the war effort in previous wars. It wasn't much different to what I heard occurred in Korea and World War I or World War II. We didn't sense the negatism that the war generated not too soon after the Tet Offensive, when it came apparent that we were going to pull out," Daly said.

"We were in a sense proud that we were going to play a role in something we felt our government deemed to be of the highest priority. Hundreds of thousands of others had preceeded us. Many had given their lives. Those of us who had trained for years for this kind of eventuality, we didn't relish the thought of being engaged in warfare, but that is what we had been trained for and felt we were carrying out the mission and the objectives of the United States government. We were proud of being given the responsibility to do that," he said.

"That pride disappeared not too soon after those years," added Daly. "I was very proud because what we found ourselves engaged in was some really deep life-or-death exercises. At the time I got over there the Battle of Khe Sanh, very soon after our arrival, took on historic proportions. There were tens of thousands of soldiers on the ground. We knew that without successful B-52 bombing intervention, they would be overrun within a matter of hours. We had a real purpose to the majority of the missions we flew. We knew that the people we were supporting were in desperate straits. They were fellow Americans and needed our help badly. We could focus on the mission of the day and we'd fly day and night, on and on without stop in support of people who were in trouble," he said.

"The routine was pretty much fly, eat, sleep, study your next mission. We were going round the clock. Often times you got into sort of a zombie type of a state because you were really doing nothing more than supporting the troops in Vietnam....If you flew from Guam, you had an 18-hour mission. If you flew from Thailand your mission could be two to three hours in length. If you happened to be in Thailand that month you'd fly two or three missions a day. If you were in Guam you would fly the most every

DALY'S CREW-- Capt. James Stewart, electronic warfare officer; Capt. Michael Daly, navigator; Capt. Reimer, RB; and Maj. Wolfgang von Kressen, pilot.

day, one and a half because of the different lengths of the missions. From Guam there was refueling involved. Over the Philippines or from Thailand there was never any refueling involved," said Daly.

"The one mission I recall quite vividly occurred January '68 when my crew received historic distinction of being the first crew to ever be fired upon by a SAM-3 missile. And while hundreds of crews had reported SAM-3s were in the area -- and we could monitor the electronic signals that they were searching for B-52s -- never had one been fired. We earned the distinction of being the first ever to be fired at," he said.

"I recall that mission like it was five minutes ago. The electronic warfare officer notified us that the missile had been launched. There had been three in tandem.

We knew within 12 seconds of the launch signal we would either be dead or alive. Basically we went into our evasive maneuver and we all counted off the 12 seconds. About the 12th second, the pilot reported that three missiles had gone by our wings. We breathed a sigh of relief and continued on," Daly said.

"We returned from that mission quite successfully and were met by virtually every intelligence official in the Air Force who wanted to recap every single instant and moment of that particular flight because it was a flight of historical proportions and set a new stage of activity for the war in the air," he said.

After a half year of 99th Bombardment Wing operations in Southeast Asia, Daly's crew looked forward to returning to Westover. "This was a time of great expectation and joy for everyone who had returned. We had been looking forward for seven months for the day we got back in the states and saw our families. It was an adventure, one that we were delighted to have behind us and just the thrill of coming back to friendly territory and family was beyond description. By that time some of the negativism of the Vietnamese effort had really come to the surface and in addition to wife, children and immediate family, there weren't too many people to see us return. It was really a family thing....At that point we were so happy to come back that we were not looking for any heroes' accolades. We just wanted to be done with that portion of our lives. I know in my own case, I left the Air Force soon after that, as I had planned, and went on to a different phase of my career and I applied my energies

WOLFGANG'S WEHRMACHT was the name of Capt. Wolfgang von Kressen's B-52. This photograph was taken when the aircraft returned to Westover in 1969.

there....We had returned to Westover without fanfare....We came back in over prolonged periods of time and there wasn't a single occasion or event that marked the return of the Westover crews. We would come a day, two days apart and the returns would stretch out over weeks and months. I don't think the community could have organized any kind of reception of note for us," said Daly.

His reaction to learning of the Eighth Air Force deactivation and the loss of the 99th Bombardment Wing was one of anger toward the administration of President Richard Nixon. "I had the impression when Westover was closed that it was done as a political backlash. As a result of the Commonwealth's support of a particular presidential candidate versus another candidate. I felt that was a terrible statement to be made by such a high office -- that in fact, we could punish communities because of their political persuasion. That angered me more than anything I can recall," he said.

"I was just a captain, so I never had access to the strategic thinking of the Joint Chiefs of Staff or the Air Force generals. I personally felt Westover had a strategic location, was the finest air base, had the capacity and runways to rival any base in the world."

Daly describes himself as "the type of individual who came out of the Sixties, who was never a career officer. My attachment stems from four years of very pleasant interaction with people like myself, people who were committed. I have very fond memories of that period in my life. Being in an Air Force B-52 crew at Westover was a very rewarding period of my life, and I think that is why I look back upon it favorably....You can't ignore a life with five or 10 individuals with the stakes so high....Where your life rode on your skills and theirs on yours. Given the fact that I am still around, I am very thankful for their skills," he said.

"Westover has a long unique history, a proud history and is something the immediate community outside the air base has a right -- and almost a responsibility -- to of which to be proud.

I would hope the same friendship and bond I encountered during the Sixties, that people who are using Westover now, will develop in the future. It is well worth the partnership that evolves."

While the Tet Offensive of 1968 caused more Americans to doubt the progress of the war in Vietnam, the invasion of Cambodia in the spring of 1970 was the watershed for public opinion. Four students were killed by Ohio National Guard troops at Kent State University and colleges across the nation went on strike to protest the widening war in Southeast Asia.

Westover AFB became the symbolic site in Western Massachusetts to protest the war. Students and other anti-war protestors gathered at Westover's gates with signs and bullhorns. One day, 700 protestors were arrested.

"Every morning we had a floor show....Lots of college students and professor types. Some barefoot, some costumed, some wore gas masks, all kinds of get ups. I saw midnight Masses on orange crates. Sixty to 100 people coming up the street with candles at midnight in the form of a cross marching in from Springfield, but every morning we had a floor show. They were no problem, only that they blocked traffic for hours, this wasn't bad because our morning business was mostly production. And if I could keep my employees on the job instead of in the windows, half of my job was done. It could be irritating at times, especially irritating if I saw them desecrating the flag. We had many instances of them coming into the store and buying things, we stopped selling them flags because they were stapeling them upside down on things. But there were times when the security of the base was threatened and I remember all of the measures and precautions that they did, and one time it was beneficial for us because they brought in the 82nd Airborne, and there was a good number of them on base, and we had fatigues to press all weekend, we were working around the clock, Saturday, Sunday cause they had hundreds if not thousands of extra fatigue uniforms that had to be washed starched and pressed by a certain time, and the first of the week they packed up and went back to wherever they came.

-- Neil O'Leary, owner Sentry Uniform & Equipment Co.

WESTOVER'S MAIN GATE was the site of many anti-war protests.

"I think one of the saddest days of my life at Westover," recalled Betty O'Connell, "was the day I heard the Eighth Air Force was leaving. I was sitting right here. I had the day off and my family was sitting here from Louisiana and the secretary to the boss called me and she said 'Guess what? Eighth Air Force is being done away with!'"

The once mighty Eighth Air Force moved out in 1970. "I have a lot of friends in Washington and when I was visiting down there, people who knew what they were talking about said, 'In the morning, word was around that Second Air Force was going to close down in Louisiana,' and by the time the politicians got through in the afternoon, it was the Eighth Air Force," said O'Connell.

The 99th Bomb Wing continued Vietnam missions through the "Christmas Bombing" of 1972 that brought Hanoi back to the bargaining table. In early 1973, American prisoners of war were released from North Vietnam and many arrived at Westover for recuperation at the base hospital. Soon after Operation Homecoming closed at Westover, word was received that the 99th Bomb Wing would also be departing the base. With the cut-backs following the Vietnam build-up cycle, Westover AFB was to enter a new era.

GOING HOME to Westover. A very thin Capt. Charles A. Brown Jr. of South Hadley, Mass., shakes hands with an American officer at Hanoi's Gia Lam Airport minutes after he was released from the "Heartbreak Hotel Annex." Brown, who later became a lieutenant colonel at Westover, was captured when his 99th Bomb Wing B-52 was struck Dec.19,1972, over Hanoi by a missile.

WESTOVER IN TRANSITION

Deactivations at Westover AFB spark civilian development of the base

By 1964 the economic impact of Westover AFB was $40 million a year in payroll alone. Local purchases averaged $1 million a month. But the major impact was cultural as Air Force families were found in civic and community organizations throughout the Connecticut River Valley, and the children of Westover officers and enlisted men were found in nearly every area classroom. Congress had passed Public Law 874 to reimburse local communities for educating children of federal employees, and Public Law 815 to grant school assistance to federally impacted areas.

As the war in Vietnam was winding down in 1972, the annual pay roll for 5,300 military and 800 civilians at Westover AFB was calculated at $54 million.

Then came the shocking news.

"I remember my husband coming home and telling me they were going to close the base. This was before it hit the newspapers, and I said, 'No! Come on. They'd never do that!' The base was too important. I guess because we lived here and because I had seen so many things over the span of time come out of there and go into there, I thought strategically, and every other which way, because it had been used in so many different ways -- during crisis, and during war, and during peace -- that it was indespensible. That there was no way they could close Westover."

-- Carol Bernier-Gawle, resident of Chicopee

In the media wake of Operation Homecoming, Westover was again in the news as U.S. Reps. Edward P. Boland and Silvio O. Conte announced on April 20, 1973, the partial closing of the base and that portions would be turned over to the Air Force Reserve. The move was part of cutbacks at 273 other federal installations as the Vietnam War entered what Dr. Henry Kissinger was later to term "a decent interval" before the Saigon government was left to its fate.

The original Defense Department plan was for the 99th Bomb Wing to be inactivated in the summer of 1974, but the event came sooner and the Air Force Reserve took over on April 1, 1974. One of the first tenant units to depart was the 819th Civil Engineering Squadron (RED HORSE) which had spent three and a half years in Vietnam at Tan Son Nhut air base. The squadron left in June for McConnell AFB in Kansas.

The 590th Air Force Band of the East ended 18 years at Westover with a farewell concert on July 12, 1973, and departed the base on Aug. 24 for McGuire AFB, N.J.

When school opened in September, enrollment in the three base schools was down by 556 students and Chicopee officials reported the loss of another 890 pupils.

At the start of 1973 there were nearly 5,000 active duty SAC members at the base. By Christmas Day there were only 1,600 SAC personnel left at Westover.

"Senator Brooke was screaming about the B-52s harming civilians, so he didn't sound like someone who could save a B-52 base....Senator Kennedy, he was also a dove and what we needed at that time were some strong hawks....you can't be asking to cut the defense budget nationally and keep it strong in Massachusetts. You just can't do those things. We didn't have the cards."

-- Neil O'Leary

One way to keep Westover AFB as a national asset and available to the Department of Defense in time of emergency was to turn control of the base to the Air Force Reserve. Traditionally, Reserve and Air National Guard units were stationed at civilian or military airports in a small cantonment area, so the decision to have the Air Force Reserve run a major military air base was a significant advancement in Pentagon thinking and a tremendous vote of confidence in the role of the citizen-airman.

As the Air Force Reserve took control of Westover AFB, a new quasi-governmental agency was established to make the surplus acreage and buildings abandoned by SAC into a profitable enterprise for the surrounding region.

Westover Metropolitan Development Corporation was incorporated by a special act of the Massachusetts legislature in

"I remember there were military and retired military people saying, 'They'll never close this base; they'll bring something in.' These were soothing words they were throwing at us....The most favored story and the biggest rumor was that the Air Force was going to close MAC at McGuire in New Jersey and move it back here where it once was. And plenty of ex-Air Force people would say, 'They're crowded down there. Jamming the New York air traffic area. They're dying to come back here. That's what this is all about.' And it went on and on....Every time a new person was on station or a new airplane arrived at Westover with officials of one type or another, a new story would come out. 'They're up here checking the runways for a new mission.' That was a real feast of rumors."

--- Neil O'Leary

WESTOVER INDUSTRIAL AIRPARK was created by WMDC at the southern end of the air base. In the foreground is the "Christmas Tree" where nuclear-armed B-52s were on alert for nearly two decades of the Cold War. In 1988, "the Molehole" was transformed into a civilian air passenger terminal. (photo courtesy of WMDC)

> **We were blessed. What the Air Force did was hold all that land and give it back to us at a time when it's worth thousands of times what it was.**
> **--- Neil O'Leary**

1974 and began planning for what would become in 15 years the catalyst for $350 million in economic and industrial growth. WMDC established three airpark industrial areas and a municipal airport operation on former Air Force property.

Citing the key location of Westover AFB to the major interstate highway intersection of I-91 and I-90, the Massachusetts Turnpike, and the proximity of I-391 and I-291, WMDC representatives touted the industrial park as being at the crossroads of New England. They pointed out that Westover was within a day's drive of more than 30 percent of the U.S. consumer market.

The civilian aviation potential of Westover's wide and long main runway led in 1981 to the creation of Westover Metropolitan Airport and the basing of air freight operations to Europe by Emery Worldwide through Chicopee Air Cargo. The airport was licensed by the Massachusetts Aeronautical Commission and a terminal was created above the "Molehole" alongside the "Christmas Tree."

Stony Brook Air Force Station became the Stony Brook Industrial Area of Air Park East. Massachusetts Municipal Wholesale Electric Co. erected a huge power plant in the former weapons area. An explosives firm, Ensign Bickford Co., made use of the bunker area while seven other firms moved onto other parcels within the

MMWEC POWER PLANT was built by the Massachusetts Municipal Wholesale Electric Company on the former nuclear storage facility at Stony Brook Air Force Station. (photo courtesy of WMDC)

Ludlow portion of the base. Sites were occupied along Perimeter Road by Advanced Drainage Systems Inc., a pipe manufacturer; Cindex Industries Inc., a tool manufacturer and remanufacturer; Custom Machining Industries, a tool and die works; J & S Tool and Die Co.; Worthy Tool & Die Co.; Manchester Electric, a contractor; and Stuart Sport Specialties Inc., a sporting goods manufacturer.

The balance of Airpark East was also developed in Ludlow on 180 acres east of MMWEC's 511-megawatt facility and south of Westover Golf Course, which was taken over by Ludlow. Fourteen firms, employing nearly 600 workers were based there in 1990 on 47 acres. Airpark East firms included ALCA Distributing Co. and A.J. Sleeper Co., distributors of health and beauty aids; American Relief Abroad, a Polish relief organization; Bay State Supply and Service, a propane gas supply and service firm; Barber Associates, a distributor of overhead and industrial fire doors; Bassett Boat Co.; Butler Corp., a specialty van conversion firm; City Stamp Works Inc., designers of flexible and steel dies; East Coast Industries, designers and installers of restaurant interiors; Kleeberg Sheet Metal Inc., fabricators of metal products; Lonner Industries, manufacturer of specialty products; Medical Management of New England, a medical billing service firm; Roma Marble Inc., manufacturers of bathroom furnishings; Stan-Mar Vendors Inc; and Vallee Heating and Air Conditioning, a contrac tor.

Airpark West was created alongside the Industrial Gate area and 30 firms employing 1,650 workers developed on the 180-acre parcel. Some of the major firms were Chemlawn, a lawn care service; Sweeney Brothers Inc., trucking and storage; American Specialty Grinding Co. and American Tool Supply Co.; Burke Beverages Co., regional distributor for Coors beer and Guinness; Commonwealth Packaging Corp.; Fairview Machine & Tool Co. and National Automated Systems; Forging Tool & Die Co.; Kraft/S.S. Pierce, a restaurant supply company; Ludlow Technical Papers, a specialty paper manufacturer; Lussier Industries, a construction firm; New England CRINC, processor of recyclable materials; North Atlantic Millwork Co., manufacturer of NAMCO windows and doors; Poly-Plating Inc., a metal plating firm; Poly-testing Inc., a metal testing firm; Popular Metal Products Inc., a sheet metal firm; Quabbin Industries Inc., a specialized machine shop; Silver Spoon Day Care; Sonoco Products Co., a paper product firm; Springfield Corrugated Box Inc., manufacturer of shipping supplies; Standard Manufacturing Inc., makers of photographic machinery; Stenor Tool Co.; Stop & Shop Companies, supermarket distributors; Sundor Brands Inc., manufacturer of fruit juice products; United Liquors West, a distributor; Westover Color & Chemical/Advanced Resins, a compounding facility for specialty plastics; and Bob Wiener Tire Co., a recapping firm.

Airpark North was established in 1989 on the former Duncan Knoll housing area. The homes were demolished and Dennisson National paper company began construction the following year to move from Holyoke to Westover with about 1,500 jobs.

MAJOR DAVID T. MOORE of Westover's 337th Tactical Airlift Squadron pilots a C-130E of the 439th Tactical Airlift Wing on a low-level mission over Western Massachusetts. (Faulkner photo)

THE NATION'S LARGEST RESERVE BASE

Post-Vietnam policy makes Westover an important ingredient in the 'Total Force'

A new era dawned at Westover AFB on April Fool's Day in 1974 when the 439th Tactical Airlift Wing was activated and took control of the base.

The 439th TAW had an odd heritage, one which probably fit well with the newly assigned role as the inheritor of such a famous, but nearly dormant, air base and a hodge-podge of old aircraft.

Lineage of the 439th was linked to a group of C-47 pilots who flew in World War II and took tremendous casualties for crews assigned to unarmed airlift missions. The 439th Troop Carrier Group spent Christmas of 1944 aiding the "Battered Bastards of Bastogne," the Screaming Eagles of the 101st Airborne Division. By most standards, the resupply of the 101st Airborne was a successful operation, thanks to dedicated C-47 crews, brave glider pilots, and well-trained paratroop Pathfinder teams. Army estimates of the airlift operations put 90 to 95 percent of the bundle drops landing where the beseiged paratroopers could reach them. But some days there were high costs among the airlifters.

One such day for the 439th Troop Carrier Group was Dec. 27, 1944. Six days before Christmas, a German raiding party overran an entire medical company. Only a few surgeons escaped. The day before Christmas a light plane landed a surgical team and some medical supplies. Messages came from Bastogne over Christmas requesting medical volunteers to parachute into the surrounded city to care for the wounded and dying paratroopers. On Dec. 26, 11 gliders were towed into Bastogne and landed successfully with surgical teams and 600 jerrycans of gasoline.

The following day, another medical mission was launched, but this one was massive compared to the day before. The 439th Troop Carrier Group towed 50 Waco gliders behind their C-47s. Only 35 made landings near Bastogne. German gunners shot down 15 gliders, killing or capturing the glider pilots, and the 439th TCG had 13 C-47s shot down.

After the war, the 439th Troop Carrier Wing was formed as an Air Force Reserve unit at Selfridge AFB, Mich., on June 27, 1949, to operate C-46 *Commandos.* A year later the Korean War began and on April 1, 1951, the 439th TCW was called to active duty as were two dozen other Air Force Reserve flying wings. But two days later, like many other wings, the 439th was inactivated and personnel were reassigned individually to fill vacancies in other Air Force units. Early in 1952 the unit was redesignated the 439th Fighter-bomber Wing and on June 15, 1952, the wing was activated again at Selfridge AFB. The 439th FBW flew T-33s, the two-place trainer version of the F-80 *Shooting Star*, and F-84E *Thunderjets* until deactivation on Nov. 16, 1957. The 439th was deactivated because the aviation reserve fighter role was placed in the Air National Guard.

The colors of the 439th were put in storage until March 1974 when the wing was redesignated the 439th Tactical Airlift Wing and activated April 1, 1974, at Westover AFB and placed under the control of the Tactical Air Command. By the end of the year, the 439th TAW was shifted from TAC to the Military Airlift Command with the Twenty-first Air Force at McGuire AFB as the intermediate gaining command.

Given new life in 1974, the 439th TAW was also given a hefty responsibility in running Westover AFB and two flying squadrons -- the 337th TAS and the 731st TAS.

The 337th had been with the 905th Tactical Airlift Group formed in 1958 at Bradley Field with C-119s *Flying Boxcars.* When the 337th converted to large C-124 *Globemaster IIs* in 1965, the whole 905th TAG moved to Westover. In 1972 the 905th TAG traded in the C-124s for C-130B *Hercules.* With the activation of the 439th, the 905th was deactivated leaving administrative, maintenance and other personnel to be absorbed by the new wing at Westover.

The other flying squadron, the 731st TAS operating C-123K *Providers*, was formerly assigned to the 901st TAG at Hanscom AFB. The 901st Came to Westover in October 1973 and was also deactivated in March 1974 when its personnel were brought into the latest incarnation of the 439th.

Although the whole activation of the 439th TAW seemed a bureaucratic paper shuffle -- with former 901st people supporting the 731st and former 905th people supporting the 337th -- the "hard luck" 439th wing had a chance to grow and to find its own way with new faces and facing the challenge of running the nation's largest Air Force Reserve base.

The assignment called for a tough commander, and the Air Force Reserve selected an old-time warrior, Billy Knowles of Goose Creek, Texas, who had been commissioned 31 years earlier as an 18-year-old pilot. In addition to his European Campaign Ribbon with three battle stars, Brigadier General Knowles wore the Air Medal with five clusters and the Legion of Merit.

Brig. Gen. Billy Knowles

Knowles ran Westover and the 439th TAW until 1978 when he was named commander of the 94th TAW at Dobbins AFB, Ga. He retired from the Air Force Reserve in 1981. During his military career, General Knowles flew nearly 8,000 hours in 16 different aircraft. He retired in March 1990 from his nine-year civilian post as director of operational plans for the Air Force Reserve.

A later commanding general of the 439th termed Knowles "a warrior who certainly left his mark on the entire Air Force Reserve. He was well known as a straight-forward dedicated officer who always accomplished the mission."

The operation of Westover AFB and Air Force Reserve units in the mid-1970s called for a tough commander like Knowles. The Vietnam War was in what Dr. Henry Kissinger would later term the "decent interval" stage between the U.S. troop withdrawal in 1973 and the fall of South Vietnam, Cambodia and Laos in 1975. In those years the Air Force Reserve was still a haven for young men who had sought an alternative to "greetings" from the U.S. Selective Service System and harbored few military career goals other than discharge. Until enlistment terms expired in 1979, the units at Westover were plagued with some Vietnam-era enlistees who had escaped the draft.

But the end of conscription in early 1973 was also having a profound effect on the whole system of reserve military forces in the United States. With the end of the draft and creation of the new "volunteer army," the nation's National Guard and Reserve system needed rapid revamping to make part-time military service an attractive avocation for those in uniform and for their civilian employers. Westover AFB was to find itself on the frontline in the new image and operational status of the Air Force Reserve in the post-draft era.

The answer was wrapped up in the "Total Force" package, the concept of a smaller active duty military supported by a much larger and more professional reserve component equipped with state-of-the-art equipment and flying machines. The new "Total Force" structure was also politically expedient because the new system effectively blocked the U.S. from conducting future Vietnams without calling up the Reserve and National Guard forces, an ideal which appealed to both the public and the Congress.

At Westover AFB the new look of the Air Force Reserve was manifested by the rapid growth of Billy Knowles' 439th Tactical Airlift Wing.

On Jan. 25, 1976, Knowles gained control of the 914th Tactical Airlift Group at Niagara Falls, N.Y. Like the 439th, the 914TAG and its flying squadron, the 328th TAS, had undergone many incarnations and counted many veterans within its ranks. The 914th TAG traces its lineage back to the First Combat Cargo Group, a China-Burma-India unit activated in 1944 for operations flying the Hump. The unit's flying squadron -- once the 3rd Combat Cargo Squadron (C-47s), and later the 328th Troop Carrier Squadron (C-46s) -- had flown C-119s since 1957 and the 328th had been activated during the Cuban missile crisis in 1962. By the time

439th TACTICAL AIRLIFT WING C-130 Hercules on the North Ramp at Westover AFB.

Knowles had control, the unit had been flying C-130As for five years.

In 1977, the 439th received the Air Force Outstanding Unit Award for providing 90 percent of the worldwide air support needed by the U.S. Army's 10th Special Forces.

The wing grew again on Oct. 1, 1980, when the 911th Tactical Airlift Group at Pittsburgh was transferred to the 439th TAW. The 911th began in 1963 as the 911th Troop Carrier Group with the 758th Troop Carrier Squadron flying C-119s. In 1966, the 758th converted to C-124 *Globemaster IIs*, and then to C-123Ks in 1972. The 758th was retrained in 1980 from C-123s to eight C-130As. Seven years later, the unit received eight factory-fresh C-130Hs.

By 1981, Air Force Reserve members of the 439th Tactical Airlift Wing were operating 41 tactical airlifters -- 16 C-123Ks and 25 C-130A/Bs.

Sixteen C-123Ks and nine C-130Bs at Westover were exchanged in 1981 and 1982 for a fleet of 16 C-130Es. A series of unit deactivations and activations were accomplished to streamline the Westover operation. In essence, the 901st units were melded into the 905th and, three years later, nearly all were redesignated as 439th units.

CAMOUFLAGED C-123K of the 731st Tactical Airlift Squadron climbing out of Westover AFB. Sixteen C-123 Providers were stationed at the base from 1973 until 1982.

In August 1981, Westover AFB hosted *Condor Redoubt*, the most massive Air Force Reserve exercise ever held. More than 3,000 reservists and 200 aircraft participated in the two-week training mission. The 439th TAW wing commander, Brig. Gen. Donald E. Haugen, served as commander of the 414th Composite Wing, created to conduct the training exercise. The major event of the two weeks took place on "Surge Sunday," Aug. 23, 1981, when normally quiet Westover was transformed for four hours into a major operational area for paratroop drops, strafing and bombing runs, low passes by gunships, short-field assault landings and medical evacuations. Although Westover was briefly in the local press for *Condor Redoubt*, the operation was a significant execption to the local perception that the base was barely active.

The C-130 era at Westover was in its heyday between 1982 and 1987. With 16 well-maintained E-model *Hercules* stationed at the base and the consolidations of the various squadrons completed, Westover reservists settled in for five years to a normal training sequence of one weekend a month and two weeks a year. There were approximately 400 Air Reserve Technicians and 500 civilian workers at the base to maintain both the facilities on Westover and to maintain the aircraft.

Medical personnel of the 74th Aeromedical Evacuation Squadron trained to prepare patients for emergency medical airlift from a battle zone. The 42nd, 58th and 59th Aerial Port Squadrons received practical training preparing cargo, loading aircraft, and rigging parachutes for massive air drops of large and heavy equipment, CDS bundles and LAPES loads. The 439th Civil Engineering Squadron and other Westover AFB units deployed frequently to other bases and overseas for training. In short, Westover AFB while under the control of the Air Force Reserve was

LOW-ALTITUDE PARACHUTE EXTRACTION SYSTEM was developed in the early Sixties and proved valuable, as shown in the photo above at Khe Sanh. The 439th TAS was frequently tasked to airdrop experimental delivery systems under development by the Natick Laboratories.

functioning as an excellent training facility for the "Total Force" mission.

Practical aircrew training was achieved by flying the low-level routes around New England. Although the tactical airdrop missions were physically grueling for aircrews, they added excitement to the lives of reservists who flew airliners or worked in New England as school teachers, police officers or any one of a hundred civilian occupations. Training missions to Aviano Air Base in Italy were also common.

Each year since 1978, the 337th MAS and the 439th Organizational Maintenance Squadron deployed aircraft and reservists for a month to Howard Air Base in Panama for Volent Oak missions around Central and South America. Along with gaining valuable training by flying in unfamiliar terrain and under different atmospheric conditions, the 337th usually managed to return to Westover with accolades for accomplishing some emergency action.

In June 1981, the seven-man crew of a Panamanian freighter, the *Kanaloa,* radioed an SOS when the ship was listing to port and sinking. A Westover aircrew led by Capt. B.J. Ferro III circled the ship for five hours until a Columbian tugboat arrived from Baraquilla.

More than 150 American civilians were evacuated from Bolivia during a coup attempt in November 1979. The following

CONTAINER DELIVERY SYSTEM gave the 439th TAW C-130s the ability to rapidly resupply ground troops in combat. Low-level training routes were developed in New England so Westover's airlifters could simulate wartime operations.

year, 439th aircraft participated in REFORGER (REturn of FORces to GERmany).

Crews from the 337th TAS also participated in Red Flag exercises at Nellis AFB, Nev., and learned to survive in an environment filled with MiGs, groundfire and heat-seeking missiles. Westover aircraft and crews went yearly to Volant Rodeo, an international airlift competition of the Military Airlift Command, and usually returned with trophies for accuracy and professionalism.

HEAVY DROP by huge cargo parachutes from C-130E Hercules.

Upon returning from the Volant Rodeo in June 1984, 439th TAW wing commander, Col. Jack P. Furguson, told the city editor of the Springfield *Daily News* he was "encouraged and optimistic" that huge C-5A Galaxy transports would be stationed at Westover. Ferguson, a former C-5A navigator, said, "The Air Force is worried about the weather in New England, but Westover AFB would be a great jumping-off point for Europe."

Earlier in 1984, Chicopee Mayor Richard S. Lak said the base had been passed over and the first Air Force Reserve C-5A wing would be based in 1986 at Kelly AFB in Texas.

MISS PIGGY* was created by the Westover AFB Fire Department to train base firefighters to battle 2,000-degree fuel fires. On April 10, 1979, an acetone and alcohol fire erupted at Hampden Color and Chemical Co. in Springfield's "Gasoline Alley." The three-alarm fire burned out of control for hours until Westover firefighters were called. Within 40 minutes after the call was received at the base, a Westover crew with a single 550-gallon P-4 foam truck had the fire under control. Today, firefighters from the communities surrounding the base are trained at Miss Piggy to familarize officers with base firefighting capabilities.

OPENED WIDE to receive a UH-1D Huey is a C-5A Galaxy of the 439th Military Airlift Wing at Westover AFB.(Faulkner photo)

STRATEGIC AIRLIFT MISSION

Controversy over loud C-5As renews interest in the base

The optimism of Jack Ferguson, who later became a brigadier general with the 439th Tactical Airlift Wing, was well placed. Local elected and business officials lobbied hard to have C-5As assigned to Westover AFB. In January 1985, U.S. Rep. Edward P. Boland announced the Air Force decision to base the giant *Galaxies* at Westover. At first, the decision was for eight C-5As, but the number soon grew to 16 aircraft.

An environmental scoping meeting was held in September 1985 to solicit the concerns of Westover's neighbors. Questions were raised about any effect on the environment, but most local residents attending the scoping hearing at a Chicopee school auditorium were concerned with the noise of the four huge turbofan engines.

While the Air Force Reserve was preparing an Environmental Impact Statement, the Westover Metropolitan Development Corp., which had gained access to the base runway in 1981, asked the Air Force for permission to have 24-hour operations at the base. In July 1986 the expected release of the EIS was delayed to study the WMDC proposal. The EIS was released at the end of the year and Westover was in the news daily as some residents of Granby and other towns surrounding the base mounted an organized campaign to stop the C-5As from coming to the base.

Following a well-attended public hearing in December 1986, the Air Force allowed an extended public comment period before releasing the final EIS in April 1987. The following month, the official decision was made to station 16 C-5As at the base beginning with the next federal fiscal year in October 1987. No decision was made on the WMDC proposal for around the clock airfield operations.

More than $50 million in construction was soon underway to accommodate the fleet of strategic airlifters. New faces appeared at the base as the Air Reserve Technician and Civil Service forces were increased to handle the new aircraft and mission.

One protest group attempted to delay the basing of the C-5As by filing two suits in federal court, but both suits were dismissed.

The controversy over the C-5As caused many residents of Western Massachusetts to renew their ties with Westover. Base personnel received letters and telephone calls of support from people who had worked or had been stationed at the base, been born at the Westover hospital, or were somehow affected by Westover. A Chicopee teacher, John D. Gordon, began researching and filming a video documentary on the base.

The basing of 16 C-5A *Galaxies* at Westover also became a catalyst for other development in the Pioneer Valley. Local communities which lacked zoning regulations or building codes adequate for sites near an airport began taking the development of the base seriously. The Pioneer Valley

THE LAST WESTOVER C-130 is towed from the Base Hangar as Brig. Gen. Frederick D. " Mike" Walker, the 439th commander from 1986 to 1990, walks alongside.

Planning Commission enjoyed community support when a Joint Land-Use Study was designed to take advantage of the Air Force's routine Air Installation Compatible Use Zone study.

The economic impact of the base soared. Before the C-5As arrived, the base spent about $40 million a year on payrolls and local purchases. With the expansion of the work force and necessary construction for the C-5As, the base reported spending $110.4 million in 1989.

Ramp construction at Westover, 1988.

During the first year -- despite the limited number of C-5As assigned to the base -- citizen-airmen from the 439th Military Airlift Wing flew more than 1,900 hours and carried military cargo and passengers to 20 nations in the Pacific, Europe and the Middle East.

Westover's C-5A crews also transported President Ronald Reagan's helicopter and limousine to Belgium, Canadian forces and vehicles assigned to the United Nations Peacekeeping Force for the Iran-Iraq ceasefire to Turkey, volunteers and vehicles to fight forest fires at Yellowstone National Park, Navy supplies to the Persian Gulf, equipment to the Philippines, relief supplies to Honduras, and military cargo to Guam.

Members of the 439th MAW flew aboard C-5A *Galaxies* that first year to Hoduf and Dahrain, Saudi Arabia; Manama, Bahrain; Ineh'on, Korea; Tel Aviv, Israel; Adana, Turkey; Yakota, Japan; Cabanatuan, Philippines; Cantania, Trieste,

Avanio and Sicily, Italy; Cadiz, Rota and Madrid, Spain; Sondre Stromfjord, Greenland; Keflavik, Iceland; Frankfurt and Ramstein, Germany; Guam; Brussels, Belgium; Rotterdam and Amsterdam, Holland; Mildenhall, England; Bermuda; Honduras; Panama; San Juan, Puerto Rico; Honolulu, Hawaii; Anchorage and Fairbanks, Alaska; and in Canada to Trenton and Toronto, Ontario; Halifax and Yarmouth, Nova Scotia; St. John, New Brunswick; Goose Bay, Labrador; Alberta, and Quebec.

Across the country that year, Westover's *Galaxies* flew missions to Delaware, Florida, North and South Carolina, Georgia, Virginia, Pennsylvania, New York, Ohio, Indiana, Michigan, Mississippi, Oklahoma, Nebraska, Wisconsin, Montana, California, Oregon, Washington, New Mexico, Rhode Island, New Hampshire, and Texas.

WESTOVER GALAXY departs Bermuda after carrying Navy Reserve Seabees to the island.

Thousands of pounds of materials collected in Massachusetts by the St. Vincent de Paul Society were airlifted Mar. 7, 1988, from Westover to New Jersey for transport to Honduras. Four thousand pounds of donated clothing from Fall River and New Bedford was placed on metal pallets and rolled aboard a C-5A and flown to McGuire AFB near Trenton, N.J.

"We off-loaded the supplies from the Catholic relief agencies and took on 120,000 pounds of other humanitarian supplies bound for Jamaica," said Lt. Col. James P. Gallin. "A C-141 *Starlifter* brought the Catholic relief materials to Honduras."

In 1988, Air Force Reserve members and the giant C-5As at Westover were tapped frequently by the State Department to perform humanitarian relief missions for Honduras, Jamaica and Armenia, plus transporting firefighting equipment during the summer blaze at Yellowstone National Park. Under the Denton Amendment, military aircraft can be used for transporting humanitarian aid and goods donated to non-profit agencies.

"These missions work well for the men and women at Westover," said Colonel Thomas G. Hargis, base commander. "Our crews obtain training and also the satisfaction of being able to help people who are truly appreciative of their efforts."

He said Westover's 16 C-5As are now flying 4,400 hours a year. "That works out to slightly more than a half year in the air," he said. "And we do this with Air Force Reserve crew members who have civilian jobs, and most have families. There is an enormous amount of dedication and personal sacrifice factored into our operations."

MSGT. GARTH PARKER of the 337th MAS guides Northeast Utilities vehicles off a Westover C-5A in Jamaica after that island was devastated in September 1988 by Hurricane Gilbert.

A C-5A *Galaxy* of the 439th Military Airlift Wing departed Westover AFB on Feb. 6, 1989, for McGuire AFB, N.J., to transport relief supplies destined for Yerevan in the Soviet republic of Armenia.

A 13-man Air Force Reserve crew commanded by Maj. David Doyle carried approximately 140,000 pounds of donated food, clothing and other supplies from McGuire AFB to Ramstein AB in Germany and then on to Incirlik AB in southeastern Turkey near the Syrian border.

The Westover aircraft arrived the next day at Incirlik AB in Adana near the Syrian border, where the supplies were off-loaded and prepared for three C-141 *Starlifter* flights into the Soviet Union. The five-day, 5,750-mile mission required 28 hours of flight.

Military Airlift Command aircraft combined two dozen missions to bring more than 700,000 pounds of humanitarian relief equipment to Armenia after the Dec. 7, 1988, earthquake killed an estimated 25,000 people and left a half-million homeless.

The C-5A *Galaxy*, dubbed "The Patriot," carried donated materials collected by the Armenian General Benevolent Union of New York. The relief materials shipped aboard the Westover aircraft included new clothes donated by the Apparel Foundation -- the charitable arm of the Clothing Manufacturers Association -- and a quarter million containers of Lipton soup. The C-5A cargo also included batteries donated by Kodak, flashlights from Duracell and candles donated by Colonial Candle Co.

When Hurricane Hugo struck the Caribbean and southeastern coast of the U.S. in September 1989, the 439th MAW had *Galaxies* winging to the three hardest hit areas within 36 hours. Westover C-5A crews brought dump trucks and heavy equipment to Charleston, S.C.; 197,000 pounds of food to Puerto Rico; and medical supplies and utility trucks from Alabama Light & Power to St. Croix in the U.S. Virgin Islands. The first Westover C-5A into St. Croix evacuated 61 civilians and carried them to MacDill AFB in Florida.

Not all missions in 1989 were humanitarian. The 337th MAS flew USMC armored vehicles to Panama in May after strongman Manuel Noriega escalated his opposition to the U.S. and refused to step down following the national elections.

The real test of the Air Force Reserve concept at Westover came just before Christmas 1989 when the U.S. launched a 200-plus aircraft operation against Panama.

Less than 24 hours after military operations began in Panama during the early hours of Dec. 20, Westover launched the first of eight missions to Central American nation. The invasion, dubbed "Operation Just Cause," came as a final option taken by the Bush administration to restore democracy, to protect American citizens, to secure U.S. interests under the Canal Treaty, and to oust Noriega.

The Panama missions were a test of both the Reserve airlift concept and Westover AFB as a major American center for strategic airlift. Development of lessons learned flying "the Hump" and from operating bases in the Assam Valley were coupled with the experiences of the Berlin Airlift, Vietnam and Grenada. Added to these operational ideas were the practical management of the Reserve which was so painstakingly developed from the late Forties, through Korea and Vietnam, and into the present concept of Westover AFB as a national asset run by citizen-airmen.

The first Westover flight into Panama was piloted by Lt. Col. James Gallin, commander of the 337th Military Airlift Squadron. The C-5 encountered several hazards when landing in the Latin American country. A major air traffic control center had been shut down an hour after hostilities began. "We flew for about 150 miles without air traffic control and used visual flight rules," said Gallin. "Avoiding mid-air collision proved to be one of our biggest problems."

As the aircraft neared Panama, tension mounted in the cockpit. Crewmembers crouched at the windows searching the pitch-black sky for enemy fire and friendly aircraft. Foreboding messages came over the radio. "Beware of incoming fire,"

OPERATION JUST CAUSE missions by the 439th MAW brought hundreds of troops and 647 tons of cargo to Panama.

warned the command post. "Possibility of attack. Land at your own risk."

The C-5A flew into the blacked-out base without lights, as did scores of other aircraft. Pilots often took evasive action while searching for the thin line of blue lights marking Howard's runway.

As loadmasters departed the C-5, the dense, suffocating blanket of humidity and the danger of their location added to the eeriness of the atmosphere, recalled Christine Mora, a crew member on four of the eight Westover missions. "The only visible light emanated from the open doors of the C-5. Loadmasters worked hurriedly to unload the plane. They ushered the rolling stock out of the cargo compartment as the troops were led away into the darkness," she said.

When the pilots returned from the Howard command post, they described the conditions. "All vehicles are driving without lights. The personnel inside are using flashlights and little desk lights to work," said Gallin.

As the C-5 departed, the pilots had to steer the aircraft around a very narrow stretch of taxi-way. "I must warn you," said the command post. "The last C-5 to attempt this landed in the mud." Members of the crew were sent out into the darkness to guide the aircraft around obstacles.

Reservists on the first flight to Panama had experienced a loss of cabin pressure. Crew members had to wear oxygen

WESTOVER LAUNCHED eight sorties into Panama in support of Operation Just Cause. Galaxies from the 439th were active prior to the invasion transporting de-icing fluid to southern bases due to a freak cold snap which threatened the 200-aircraft invasion. *(Union-News photo)*

masks for the last leg of the trip into the country and most of the way on the journey to their stopover at McDill AFB, Fla.

While readying for the trip back to Westover in Florida, the aircrew received word that they were being diverted back to Panama. As Colonel Gallin briefed the crew just three days before Christmas, he said, "Anyone who wants to go home right now can go. I just want to know -- Are we all players?"

Everyone was a player.

"Although we worked very long hours, it was great to see all the elements of the aircrew pull together and work with such camaraderie," said MSgt. Robert Perreault, an aircraft maintenance technician with the 439th Organizational Maintenance Squadron.

The first C-5 transported 52 troops from the 60th and 16th MP Brigade, Fort Bragg, N.C., and 11 vehicles out of Pope AFB, N.C. On the second trip back to Howard AB, they carried 51 troops from the 7th Infantry (Light), Fort Ord, Calif., seven vehicles, two trailers and three pallets of batteries and ammunition out of Travis AFB, Calif. A total of 647 tons of cargo was transported by Westover reservists during seven missions between Dec. 20 and Jan. 2.

In addition to the sorties flown to Panama by Westover crews, two other missions were flown in support of the

MILITARY POLICE from the 16th MP Brigade board a Westover C-5A bound for Panama. The reflective tape tied to their left arms saved many American lives during the conflict.

pre-invasion effort. De-icing equipment from Westover and other northern bases was transported to Charleston AFB, S.C., Dover AFB, Del., and Pope AFB, N.C., because severe winter storms had endangered air operations at southern bases.

"The Air Force Reserve played a significant role in the effort," said Colonel Gallin. "This shows the Reserve is right there with everyone else getting the job done, on time, under all circumstances."

PULL-THROUGH HANGAR receives its first C-5A on April 5, 1990. The $17-million hangar was designed for rapid de-icing and maintenance of the Galaxy fleet at Westover.

On April 7, 1990, the new 10-story C-5A hangar was dedicated as the 50th anniversary of Westover was celebrated. Among the honored guests were Col. Timothy O. Westover, grandson of Maj. Gen. Oscar Westover, and retired Col. Gail Halvorsen who earned fame during the Berlin Airlift as "Uncle Wiggly Wings" and the "Candy Bomber" of "Little Vittles" fame. A military ball was held that evening in the Base Hangar for 1,100 guests.

In the summer of 1990, Westover Air Force Base hosted a three-day international airshow to mark the golden anniversary of the base.